T0245570

PRAISE FOR *TRADITION MEETS TRANSFORMATION*

"No secret—North American manufacturing is at dire risk. Laurie Harbour and Scott Walton attack this threat head-on in *Tradition Meets Transformation*. At last, the roadmap to navigate from where you are to where you need to be! For business owners and leaders, your opportunity is to now take action!"

—Jack Daly, CEO, Coach, and
World Recognized Serial Entrepreneur

"Using their many years of experience, Laurie and Scott provide the tools and processes to ensure your organization is addressing the rapidly changing—and increasingly competitive—twenty-first century manufacturing industry. Their knowledge is backed by real-life anecdotes and success stories, which demonstrate how to implement their methodologies to create clear improvements and next-generation leadership for your entire manufacturing organization."

—Julie Fream, President and CEO,
MEMA Original Equipment Suppliers

"I have known Laurie Harbour and Scott Walton for years and value the partnership and friendship we have formed. They are the best in the business for manufacturers to glean insights from. As the industry looks to attract next-generation leaders, this must-read book combines Laurie and Scott's expert knowledge with modern strategies for manufacturing leaders to drive the industry into the future."

—Dave Klotz, President,
Precision Metalforming Association

"Laurie and the Harbour Results team are keenly focused on improving North American manufacturing. Continuous improvement isn't an option; it is table stakes for those with the desire to survive in manufacturing."

—Mike Meyer, President, NADCA

"Laurie and Scott have combined theory *and* practice into a meaningful guide for manufacturing leaders."

—Noel Ginsburg, Founder and CEO, CareerWise Colorado and Founder and Chairman, Intertech Holdings LLC

"In *Tradition Meets Transformation: Leadership Strategies to Revitalize Manufacturing*, Scott and Laurie capture the essence of adaptability and growth strategies for manufacturing executives in a business landscape that will continue to rapidly change. With the guideposts of years of experience and analysis, this book serves as a valuable compass, provides practical guidance, and illuminates the path to sustainable success in a fast-changing world. A must-read for those seeking to understand leadership, leverage emerging methodologies, and position their businesses at the forefront of the future."

—Troy Nix, Business Leader, Author of *Eternal Impact: Inspire Greatness In Yourself and Others*, Motivational Speaker, and Military Veteran

"Laurie and Scott are subject matter manufacturing experts. I have had the privilege of being both a client and a colleague. If you are looking for a road map on how to fix a plant, or more importantly being the leader who drives the change, then this is the book for you."

—Michael A Haughey, President and CEO, North American Stamping Group

TRADITION
MEETS
TRANSFORMATION

LAURIE HARBOUR + SCOTT WALTON

TRADITION MEETS TRANSFORMATION

LEADERSHIP
STRATEGIES TO
REVITALIZE
MANUFACTURING

Forbes | Books

Copyright © 2024 by Laurie Harbour & Scott Walton.

All rights reserved. No part of this book may be used or reproduced in any manner whatsoever without prior written consent of the author, except as provided by the United States of America copyright law.

Published by Forbes Books, Charleston, South Carolina.
Member of Advantage Media.

Forbes Books is a registered trademark, and the Forbes Books colophon is a trademark of Forbes Media, LLC.

Printed in the United States of America.

10 9 8 7 6 5 4 3 2 1

ISBN: 979-8-88750-136-9 (Hardcover)
ISBN: 979-8-88750-138-3 (eBook)

LCCN: 2023920935

Cover design by Matthew Morse.
Layout design by Lance Buckley.

This custom publication is intended to provide accurate information and the opinions of the author in regard to the subject matter covered. It is sold with the understanding that the publisher, Forbes Books, is not engaged in rendering legal, financial, or professional services of any kind. If legal advice or other expert assistance is required, the reader is advised to seek the services of a competent professional.

Since 1917, Forbes has remained steadfast in its mission to serve as the defining voice of entrepreneurial capitalism. Forbes Books, launched in 2016 through a partnership with Advantage Media, furthers that aim by helping business and thought leaders bring their stories, passion, and knowledge to the forefront in custom books. Opinions expressed by Forbes Books authors are their own. To be considered for publication, please visit **books.Forbes.com**.

To Jim Harbour, our founder, for imparting your vision to make an impact in North American manufacturing.

—Scott Walton and Laurie Harbour

To my business partner and work husband. You taught me so much about manufacturing and have shown me how to affect change. It has been a journey worth every step!

—Laurie Harbour

To Laurie, my friend, colleague, and work wife. Thanks for bringing me into the professional services space and helping me to become a better advisor and consultant. I could not ask for a better business partner or a better journey!

—Scott Walton

To our clients and colleagues and all those who let us in their manufacturing doors and who have taught us so much about manufacturing and how to make an impact.

To the entire Harbour Results Team. Thanks for your collective wisdom and for keeping us grounded to our company values.

—Scott Walton and Laurie Harbour

CONTENTS

INTRODUCTION

How did we get here?

It seems a simple question, but the answer is anything but....
It's not only complicated, but it's also evolving even as we write this book.

American manufacturing, particularly the small-to-medium manufacturing companies that built our middle class, is no longer a level playing field. Over the last several decades, an ever-changing global business environment, impacted by geopolitical, demographic, and technological influences, has backed us into a corner of our own making. For years, our advantage has been that we've been the productivity leaders in the manufacturing space, but now we're slipping. We haven't kept up with the accelerating pace of change and we were slow to recognize global competitive challenges as they arose. We've done a less than stellar job of attracting new talent into the industry. The manufacturing companies that have thrived have done so against the odds. And now, if small-to-medium manufacturers across America hope to compete, the entire sector needs to take a hard look at how business-as-usual has changed and will continue to change. And consider what we *should* do to meet the challenge.

So how did we get *here*?

Let's look at how it used to be. At the close of the last century, North American manufacturing industry hadn't changed much in fifty to sixty years other than the introduction of more efficient factory design and machinery and new manufacturing methodologies—many of them imported from Germany and Japan. The business climate for small-to-medium manufacturers was stable and healthy. Low-cost countries such as China were not yet the powerhouses in manufacturing that they are today. If you'd been in business for a generation or more, you were known and business came to you. You had a reputation—your company stood for something that your customers recognized and appreciated. You didn't need to quote or compete for every job. You had a set group of customers who relied on you to supply them with the manufactured parts they needed. The cost of materials was affordable and the supply chain that fed your production was predictable. Lean manufacturing reduced lead time, waste, and rework while creating more streamlined processes, efficient allocation of labor, and cleaner work environments. Skilled labor was plentiful.

A job in manufacturing *meant* something and was still considered a good life-long career choice. The stereotypical gritty, grimy, and noisy shop floor of our grandfathers' manufacturing plants was becoming a thing of the past, as technology in the form of computer-aided-manufacturing and production line robotics were eagerly adopted. Despite growing pressure from global competition, American manufacturing provided a sense of reliability—a steadiness of production, employment, and earnings. But that was *last* century.

A FASTER PACE OF CHANGE

By 2001, the small changes we were beginning to see in the sector gained momentum. Global competition became more challenging with low-cost countries (LCCs) entering the marketplace. China's price under-cutting tempted some of our Original Equipment Manufacturers (OEMs) and manufacturing of parts moved offshore, as did some of the sector's intellectual property. Manufacturing was evolving into a price game that stimulated more global competition. The cost of everything began to rise—material, labor, sales. Geopolitical turmoil threatened the supply chain. Technological advances in manufacturing equipment, business communications, and basic business processes started changing the pace, the face, and the feel of manufacturing faster than many operations could accommodate.

In the new century's first decade, the growth of social media platforms and society's reliance on the internet changed the way we sought information; processed local, national, and global news; and even related to each other. Old-school manufacturing practices were being eclipsed by faster, more efficient ways of producing parts, tools, or components. Manufacturing leadership—baby boomers already in their second, third, or fourth decade in the business—was faced with a technological landscape growing more unfamiliar by the day. An aging blue-collar workforce was naturally suspicious of any new wrinkle that could change how they did their jobs. Manufacturing, like any enterprise based on repetitive production, simply didn't adapt smoothly. Think of what it takes to retool and retrain an assembly line for a new automotive model. Now think of how a small manufacturer who can't afford to lose even a day's production might handle introducing a new and improved all-digital process to shift workers brought up in an analog world.

Our collective educational expectations changed over the last two generations, too, creating a gap in the manufacturing workforce that few candidates seemed eager to fill. Those of us who followed our fathers into manufacturing were likely encouraged by the promise of steady, long-term employment in a generations-old company, a linchpin of its surrounding community. There was no stigma attached to the skills learned in trade school or community college that qualified and certified you for that job in the plant, nor did "getting your hands dirty" mean anything other than a healthy all-American work ethic. But now secondary school guidance counselors are pointing every high schooler to college to pursue higher degrees and join the professional class. When and why did we decide that careers in manufacturing were not for our kids?

We entered a sustained period of economic expansion in 2008 that contributed to a general sense of complacency among many manufacturers. Modest growth was deemed acceptable in the face of ongoing competition in the global marketplace. But by 2019, our data showed that demand was waning and changing from high volume-low mix to low volume-high mix as consumers expected more options, more customization, and more differentiation. Gone were the Henry Ford days of "You can have any color you want, as long as it's black," as consumers moved from purchasing what was available to dictating how it should be. People were buying less, expansion was waning, the economy softened, and revenues declined. Volumes were dropping for cars, trucks, and planes, all consumer durable goods, and the ripple effect spread rapidly to the second- and third-tier manufacturers who weren't prepared to weather a recession, and who didn't have contingency plans in place to boost efficiency and maintain profitability. In fact, many small-to-medium manufacturers failed to understand why and how the global economy could affect their business, or what the long-term rate impacts of a more aggressive tariff policy could be.

Then came COVID-19. As the nation shut itself in, buying behavior all but stopped. Plants that weren't temporarily shut down went to skeleton crews and curtailed production. But that was short-lived, within four to six weeks everything changed. The pandemic was an unprecedented event that catapulted the industry into a generational shift. It was unlike any other recession or change to economy that the global workforce had ever experienced. No one knew exactly what to do, so we all had to learn together how to run our businesses. Many of us intimately involved in the manufacturing sectors thought the market would slow, but it didn't. Instead, consumers changed how we operated. We adjusted. We worked from home, ate in, and expanded our living spaces to accommodate our working needs. Durable goods demand actually rose 39 percent during the pandemic.[1] Manufacturing was strong and, as a result, put an unexpected strain on the supply base. Government funding meant that well-run manufacturers were wealthier and poorly run companies were able to stay viable.

But the pandemic had another unexpected impact. The Great Resignation, or unplanned retirements in the workforce, accelerated resulting in the loss of legacy manufacturing expertise. Manufacturers now faced a new, daunting set of realities. Businesses encountered material, labor, and customer challenges that were often unpredictable on a daily, sometimes hourly, basis. So many uncontrollable variables meant that proaction was almost impossible as the basic law of supply and demand was, in effect, flipped on its head. Where we used to understand demand and then match supply to meet it, now we were

---

1   "Manufacturers' New Orders: Durable Goods," FRED.com, accessed June 2023, https://fred.stlouisfed.org/series/DGORDER.

challenged to understand supply first and then decide how much demand that supply could realistically meet. Manufacturers were now forced to think in terms of how much product could be produced with the assets available, which, in turn, demanded reactive thinking and on-the-spot modification. It demanded flexibility.

So where are we now and what can we do?

Rising fuel prices, supply chain pressures, the ongoing impact of the COVID-19 pandemic, competitive threats from China and other LCCs, and the rising cost of wages, materials, and durable goods, as well as generational change in labor and leadership … all of these compromise our ability to conduct business-as-usual and operate as we did even a decade ago. Although some impacts can be offset with price increases and pass-throughs, those are dependent upon what the market will bear. Our business environment can now be characterized as being in motion. Everything is changing and will continue to change at an ever-increasing pace, especially as the economy creeps closer to an eventual recession likely in early 2024. Inflation has risen, and interest rates have only continued to go up frequently and significantly. The only way to meet these kinds of challenges, both known and unknown, is to change with them or, better yet, ahead of them. Many of the issues that have always impacted the sector have been magnified, as is the need for widespread transformation of the way manufacturing conducts business. The old-school perception of manufacturing is that it has always been a tough industrial sector. That's still true, but it's much more challenging now to compete and thrive with the global market throwing challenges at us on a daily basis. And unfortunately, the good old days are in the rearview mirror.

In short, American manufacturing is at a tipping point.

The answer, then, is to develop the flexibility to rapidly adapt to and accommodate change. Focus on getting your business through the lows while positioning to manage unprecedented growth and chaos due to labor and supply chain challenges. Easy for us to say, but much harder to do. If it were easy, there'd be no need for this book. Especially given the changing economy and the likelihood of recession, flexibility is critical.

SO, WHO ARE WE TO WRITE THIS BOOK?

As lifelong participants in the industry, manufacturing is in our DNA. We're passionate about helping North American manufacturing improve, recover, and regain the forward-thinking innovation and can-do attitude that once defined the industry. We want to share our hard-won insights, knowledge, and experience with the next generation of leaders. We want to attract new talent to the industry—talent with new capabilities, drive, and technological expertise to encourage growth and sustainable performance. We want to make a difference for your business and for the manufacturing industry.

Our business at Harbour Results, Inc. (HRI) is centered on that premise and fueled by that belief—that we can help you improve, grow, and sustain your manufacturing business through the coming generational change. Our company WHY is to make an impact in North American manufacturing and we are passionate about that mission. We've spent years gathering and analyzing market intelligence and business trends, evaluating companies, and developing strategies and tactics for enterprise-wide improvement from RFQ to delivery and from the loading dock to the corner office. Now the need to link the expertise of the seasoned workforce with the new generation of digitally native, technology-driven workers—the "smart" generation, named for their fluency with smart devices—is more critical than ever before. By providing this smart generation with the platform to impact manufacturing with their knowledge and capabilities, and with a combination of critical and creative thinking, we believe we can regain our competitive advantage and sustain North American manufacturing for decades to come.

In this book we'll describe the practices, methodologies, tools, and techniques we employ to assess where you are so you can comprehend and appreciate that reality with both accuracy and purpose. Then, and only then, can you begin the painstaking process of improvement and transformation, and evolve your manufacturing business and leadership to meet the future in a position of strength, sustainable performance, and profitability, and imbued with the flexibility demanded by the times we live in now and those ahead. This book is a guide to those who aspire to lead the manufacturing transformation and improve the competitiveness of their organization.

For many small-to-medium manufacturers, an honest assessment of how you stack up against the competition and the marketplace is not a comfortable process. Facing the truth never is. But it is absolutely necessary when looking to secure the future. Frankly, someone has to tell you what you need to hear, not what you want to hear. Although it's often a hard message far easier to dismiss than accept, it must be delivered to drive performance and make it right. And it's necessary for the long-term health of North American manufacturing.

Let's get started.

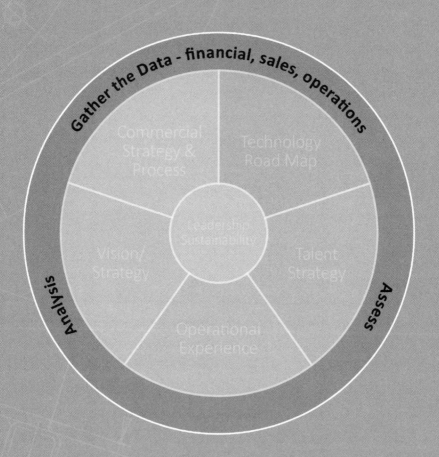

Gather the Data - financial, sales, operations

Commercial Strategy & Process

Technology Road Map

Leadership Sustainability

Vision/Strategy

Talent Strategy

Operational Experience

Analysis

Assess

IMPROVEMENT PROCESS WHEEL

CHAPTER 1

UNCOVERING
THE POSSIBILITIES

Once upon a time there was a company that started out as an automotive injection molding manufacturer but, over the years, had expanded its operations and facilities to include medical device manufacturing and brake assemblies among other things. After several decades in business, this company grew to eleven sites, including two international sites in the Dominican Republic and Canada. The problem, among many, was that there didn't seem to be much coordination or common ground between the various sites and their operations. Not only was the company's operations disconnected, but the company as a whole was financially unstable. To make matters worse, the company's leadership was autocratic and removed from the day-to-day functions of a manufacturer of its size. In effect, the company had lost its focus, its "Why." Profitability suffered. Employee morale was low. Leadership did not seem engaged. In effect, this company was a poster child for how not to grow a company.

We were retained to assess this company's overall performance and do a gap analysis—a sort of organizational health check. Our recommendations for improvement were sweeping and inclusive. We conducted many interviews, reviewed their data, unraveled a tangled history of acquisitions. In the end, we suggested divesting the business segments and facilities that were divorced from their core competencies, and refocusing on the parts of their business that were profitable. And we recommended a change in leadership.

This company fell into the matrix quadrant (discussed later in this chapter) that typifies "troubled" companies, those manufacturers in direct need of profound systemic change.

Fortunately, not all of our engagements require such drastic action.

In manufacturing, technology meets purpose. The various manufactured products, parts, and devices have a direct impact on quality of life. Whether it's a medical device, an automotive part, or a consumer product such as a plastic storage tub or a diaper changing station, manufacturing is present in more ways than most people realize. It's important to our economy, to our way of life, and to communities across the country. But in order to compete in a global economy that has become more competitive and challenged by supply chain and labor issues, North American manufacturing needs to change. Next-generation leadership will need to adopt new practices and philosophies in order to level the international playing field. And this incoming class of leadership will need to evaluate the status quo of their company, its operations, its culture, and its profitability in order to determine how to best compete in this changing landscape. They will need to define for themselves what makes a manufacturing company great for the long term. How can they sustain best practices and success?

Over the years, we've found that manufacturing companies generally fall into four quadrants of what we'll call a performance matrix. To put it broadly, the array of attributes and manifestations that populate each quadrant characterize the nature and condition of companies. These are the symptoms, good and bad, that we assess at the beginning of our engagements as manufacturing business advisors. This performance data paints the picture of a manufacturer's status quo.

We have found that many companies possess a skewed view of how they are and where they stack up against their competition, the marketplace, and sometimes even the manufacturing space in which they operate. We have a tendency to believe what makes us feel good about ourselves, our colleagues, and our companies. It's easier and far more comfortable to think about the good things happening around us, while deemphasizing or ignoring the bad. This is selective belief, a form of denial, and it's common human behavior. This is a result of not considering all the information or not believing the story that it tells. It's also the wrong way to run a business unless you want to run it into the ground. Harsh, but true.

Manufacturing Performance Matrix

Figure 1.1

As shown, there are four quadrants in our performance matrix in which the x-axis represents performance/profitability and the y-axis represents leadership/attitude. Numbered from one to four the quadrants categorize manufacturing businesses as Top Performers (1), Financially Troubled (2), Average and Willing to Improve (3), and Good but Not Sustainable (4). Obviously, the goal of any manufacturing company that falls in the second, third, or fourth quadrant is to do what is necessary to move into the top right or first quadrant. And the goal of those in the first quadrant is to keep getting better—in effect, to keep expanding and exploiting the art of the possible, which we'll talk about more as we go on.

The Top Performing companies are characterized by strong and capable leadership that continuously challenges their status quo. That is to say, they're doing well but are hungry to do even better. These are leaders who reward and acknowledge their employees' successes while pushing them to the next level of achievement. You might hear them say "Great job! Well done! Now what's next?" And rather than say, "We've always done it this way," they'd be more likely to ask, "Is there a better way to do it? " The leadership of companies in this quadrant actively seek to hire people who can bring new ideas and capabilities to their positions and to the business. For example, rather than adding yet another qualified engineer to the Engineering Department, one of our clients hired an art major because he was looking for someone who could think differently and more creatively.

Companies in this quadrant also place an emphasis on recruitment and attract talent through robust apprenticeship programs with local institutions such as trade schools, community colleges, and universities, in which candidates are placed in positions and situations where they are expected to act and solve problems rather than stand around and observe. Leaders of these companies understand innately that it is a mistake to hire in your own image and hire into their weaknesses instead of their strengths. They also hire and fire to their values, so they know that they're spending money on the right talent. They know that if they hire poorly, they're not getting the talent capability they need to sustain performance. If trying to save thousands of dollars per year per hire is their motivation, they recognize that they won't be able to retain the talent and skills they need to take their business to the next level.

Leaders of top-performing companies are aware, reflective. They know where they stack up against their competition. They benchmark and measure everything, and then react and proact to their metrics. They make a concerted effort to understand business and market trends—local, regional, national, and global—so they can be agile and flexible. They expect the unexpected. They gather market intelligence and build a presence in their competitive arena through networking at industry events. They never assume they know everything, and they never stop learning. They understand their key performance indicators (KPIs), cost drivers, core competencies, and what they're really good at, and they identify where and when they can push the boundaries of the company's performance. They embrace risk, but only if it's thoroughly analyzed and they can make calculated, informed decisions. Financially, their balance sheets are indeed "balanced," meaning that they have an appropriate level of debt to capital, are in good standing with their bank(s) and have consistent cash flow. Operationally, they have good throughput, quality, and on-time delivery—all of which contributes to double-digit profitability.

Summarized, Top Performing manufacturers are businesses that are characterized by the following:

- Profound understanding of their core competencies

- Strategically tailored to their values

- Thorough benchmarking and agile response to their metrics

- Willing to challenge their status quo and question methods, operations, and performance
- Robust financial health
- Wise cap ex and reinvestment
- Omnidirectional in communications and influence
- Well-informed but eager to learn
- Highly flexible and upwardly mobile
- Hire into diversity of age, race, and capability

At times, leaders of top-performing manufacturing companies are described as fearless. We think this is misleading. Rather than particularly courageous, we've found these individuals to have an "edge," a kind of hunger to excel that they communicate clearly across their enterprise that, in turn, influences and inspires higher performance.

Financially Troubled

The opposite of the Top Performing manufacturing company is the Poor Performing/Troubled company stuck in the lower left or #4 quadrant of the performance matrix. This type of company may be a second- or third-generation family-owned business that lacks a full perspective of their current situation or has no appreciation of how they got there. They may not realize that a positive financial statement today doesn't guarantee they won't be in trouble three months on—because they're not forward-looking and considering trending factors.

They may not fully understand the strengths and weaknesses in their balance sheet or their cash flow. Their leadership may be the only ones left in the line of family succession and may lack the skills and/or training needed to helm a manufacturing business—a de facto capability gap. They may not have learned business management the way their predecessor did, so there's a discontinuity of leadership. Or perhaps they have little interest in perpetuating the legacy and tradition.

The leader of a troubled company may equate asking for advice or help as a sign of personal weakness, and may be resistant to learning, particularly to honest assessment, and may try to "explain away" the data or discount the findings. They also are in denial; they don't benchmark anything, and in most cases they state that the benchmarking data is incorrect and that others can't possibly operate that way. They don't network or participate in industry events because they don't think it's valuable or because they feel they don't have time to be away from the plant. There will be a lack of transparency and completeness in company finances, in operational performance and labor usage, which in turn fosters less than optimal behavior across the board and fails to drive problem-solving and accountability. HR policies and practices are likely vague or muddied, and they likely have limited outside advisors and no board other than friends and family. They may feel that they don't need help when, in reality, they simply don't know what they don't know.

Poor Performing/Troubled companies are characterized by:

- Financial thinness or potentially poor business acumen

- Lack of capability

- Lack of training

- Lack of unqualified leadership

- Resistance to cultural change

- Lack of transparency/bad behaviors in finance, operations, HR, etc.

- Lack of team mentality

- Poor (know-it-all) attitude, lack of perspective

- Downward mobility

- Denial and disbelief in benchmarking or the value of networking

The company in trouble may have a tough time seeing what's in front of them because they've spent so much time and energy denying or ignoring their less-than-desirable reality. In the worst-case scenario, they may not realize that they're in trouble.

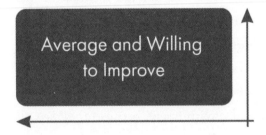

In the top left quadrant of the performance matrix, the Average Performing company likely has the right attitude insofar as wanting to improve but lacks the knowledge or know-how. They don't know where or how to start or didn't think it possible. A company in this quadrant makes money but should be making more. They're not complacent per se, but because they're not losing money they feel that they're in a good place. The owner of the Average Performing business not only takes very little out of the company, but also fails to reinvest.

The leaders of manufacturing businesses in this quadrant typically have a certain kind of humility and tend to care deeply about "their people" and their community. They lack the edginess or the hunger that drives leaders in the Top Performing quadrant to expect and demand more from their business. They always run a decent business and want to continue the tradition, but they're content with modest profitability that could be doubled with the right strategic direction and tactical tools. They accept their status quo without question. This is not to say they don't want to be better—they do! They just aren't sure how to get there. They benchmark everything but don't know how to use the information to make positive and lasting change, and they often wait for the "right" time to get an assessment or ask for help, but it never seems to be that "right" time. They attend events and network, but are never sure how to incorporate what they learn into their business. But, they keep trying!

In short, these are the companies that are making do because they don't know how to make it better. They also have great potential for improvement. They have the right people and the right culture. They have "good bones" and foundational systems that are tactically good but strategically weak. Many of the companies that fall in this quadrant struggle with accountability from the top down and tend to foster a nonconfrontational environment because they see all employees as friends and family. They equate drive and concerted action with aggression and pushiness rather than influence. In some cases, these companies are "run by the numbers," and leadership doesn't understand what is needed to grow or how to leverage growth by driving efficiency to do more with the same resources. They're not driving themselves to the next level of performance in labor utilization, technology and automation plans, sales, and marketing.

The Average Performing companies are characterized by:

- Good bones/Good structure

- Strong tactical foundation/Strategic Weakness

- Little reinvestment

- Financial mediocrity

- Contentment with the status quo

- Lack of a go-forward technology plan

- Don't know how to improve

- Business managed by the numbers

- Benchmark and network but not sure what to do with the learnings

- Market activity responsiveness that falls short of driving next-level performance

- Good attitude, but unaware of opportunity

- Struggles with accountability

- Cultural humility

- Lack of critical thinking

Although many Average Performing companies don't truly appreciate how much room they have to improve and expand before they go through the next phase of growth, they are the most rewarding to work with because they have the right attitude, leadership, and willingness to learn and excel. Once clearly communicated to them, they understand both the concept and the promise of the art of the possible.

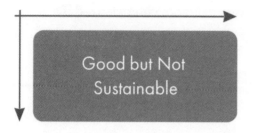

Good but Not
Sustainable

The businesses that fall in the lower right Above Average Performing quadrant of the performance matrix are the most challenging to work with because they don't believe they need help. They're performing well but hold themselves back through complacency, lack of commitment, and lack of leadership. These manufacturing companies have good market presence and reasonable to good profitability. They have good balance sheets that have created significant family wealth in legacy businesses, but earnings are lower than they should be. Often companies in this quadrant have tough cultures that don't encourage loyalty because workers don't feel empowered or engaged. Leadership/ownership also often assumes that they "know everything" and, as a result, aren't open to contributions or observations from "their" workforce. Although they invest in their business, they have a tendency to fall back into a "good enough" mindset that compromises sustainable performance and hence success. Unlike the owner of the average performing business in the upper left quadrant (#3), the owner of an above average performing business reinvests in the business, but also takes too much money out of it. In many cases these are lifestyle businesses, run by owners who use the business to maintain a lifestyle, not necessarily to drive the business to being a top performer. This group doesn't benchmark themselves often because they believe they can't possibly get any better. They believe they have nothing to learn and don't want to share how they do things because they don't want others to copy them. We sometimes refer to this section of the matrix as the "entitlement" quadrant.

Businesses in this quadrant share certain similarities with businesses in the Average quadrant in the upper left of the matrix. Many of these manufacturing companies are marked by a kind of dictator leadership style, and are hard to work for, demand a lot but fail to clearly communicate their expectations or hold their people accountable. Management structure is usually hub and spoke, where all functions and responsibilities funnel to the hub and the organization doesn't move forward unless the action is initiated at the top. That is to say that the wheel doesn't turn unless the leader decides to power it. Culturally, this is the path of least resistance. The workforce has no direct ownership in either process or success and is comfortable with the lack of accountability.

The Above Average Performing companies are characterized by:

- Strong balance sheet

- Good cash flow

- Lower earnings than they're capable of

- Hub and spoke management structure

- Workforce consistently looks for the path of least resistance

- Decision-making removed from the workforce

- Attitude of entitlement and arrogance

- Unclear expectations from leadership

- Lack of critical thinking throughout organization

- Not willing to share their data or how they do things because they consider themselves the best

The Above Average Performing company shares some of the strengths of the Top Performers such as relatively good financial practices and health but is hindered by many of the weaknesses found in average and poor-performing companies such as the lack of critical thinking and, in particular, resistance to change and learning. There's always room to improve, but if you don't consider your organization to be broken, there's little to no motivation to fix it.

A MATTER OF ASSESSMENT

Why is this performance matrix important? Because, simply, it can help you figure out where you stack up, and what you need to work on. It can help identify performance gaps, lead to the implementation of operational and financial solutions, help you develop effective strategic plans, create and implement sales processes, improve profitability, transform operations, and prioritize and address the most challenging issues immediately in front of you and on the horizon. In short, it can help you develop a road map to success and get you started on it. And to do this, you need to know exactly where and why and how you are.

So that's where you should begin. Before you examine your vision-mission-values, identify and define your company's core competencies, evaluate your competition, and analyze your throughput and any operations that impact your income statement, balance sheet, and cash flow. Step One is gathering the data that will describe *in detail* your status quo. Consider this step like your personal physical each year. We know we have things to work on but without some direction, we may not know where to start. If the doctor says you drink too much you may dismiss it, but if he or she says you have a heart blockage you may make some tough decisions to change some habits. It's the same for our business—we have to assess our current state of health.

After that, there is no concrete Step Two. Transformation is an evolving process—sometimes cyclical, rarely linear, and totally amorphous until you've completed Step One. The next step in the process could focus on any one of a half-dozen areas—operations, strategic development, commercial strategy, labor, technology, and leadership. The next step is determined by what is learned in Step One.

Once you have a clear understanding of current state, then you can start working on where to go. We call that the *art of the possible*—the strategic development of the tactical and executable plans that drive results. In other words, all the actions needed to take your manufacturing to the next level of performance. It includes leadership and cultural development, short- and long-term business plans, development and implementation of performance metrics, succession planning, labor recruitment and retainment strategy, marketing strategy and sales process development, and operations transformation and improvement. The art of the possible challenges your status quo, and it starts with your data and an honest, thorough assessment of your company.

As we said, this is not a linear process. If anything, it's more like a three-ring target or a dart board than a process diagram. The outermost ring is Assessment—the gathering and analysis of overall performance data including financial, sales, labor, and operations. Only after the Assessment phase is completed and shared can you progress to the next ring that covers all business operations including Operational Excellence, Labor/Talent Strategy, Commercial Strategy and Process, Vision/Strategy, and Technology Road map. At the heart or bull's eye of the target is Leadership/Sustainability. The Assessment phase, which is comprehensive and exhaustive, determines which section of the inner rings should be addressed first. That's why there's no set Step Two. The assessment tells us where the challenges are and how to prioritize recommended action. It guides change as we drill into specific areas of business operations.

BENCHMARKING, INFORMATION, AND EXPERIENCE

In most cases, a company's financial data tells much of the story regarding its health. When you take a hard look at the basics over a year's duration, it fixes the status quo of your company against the prevailing market conditions. You should also take into consideration market intelligence and whether you have an awareness of what's happening within the industry, including all the factors potentially influencing your numbers—such as the impact of political policy, international trade agreements, geopolitical unrest, foreign competition, supply chain pressure, unforeseen short- or long-term catastrophe (natural disasters, global pandemics) and economic expansion, contraction, inflation, or recession. What market data do you look at on an ongoing basis? How does it affect your business?

All of this overview data establishes a starting line for a comprehensive business assessment. It includes Revenue; Cost of Goods Sold (COGS); Selling, General, and Administrative costs (SG&A); Earnings Before Income Tax/Earnings Before Income Tax, Depreciation, Amortization (EBIT/EBITDA); and total number of employees (salaried, hourly, and average hours/week/employee). With this information you can set a company's business performance benchmarks for the following:

1. Cost of Goods Sold (COGS) % of Revenue

2. Selling, General & Administrative costs (SG&A) % of Revenue

3. Earnings Before Income Tax (EBIT) % of Revenue

4. Value-added Revenue/Full-Time Equivalent (FTE)

5. Revenue/Job Type

Armed with these benchmarks, you can then determine where your company appears to rank within its competitive arena, and you can ask yourself, "Are we as efficient or profitable as our peers?" "What markets are doing well and are we in those markets?" "Are others doing better than we are in similar markets? If so, why?" Without benchmarks, you don't have a basis for comparison.

SO MANY QUESTIONS

An assessment could take many shapes, but it is critical that you embark on a process that challenges you as a leader. An effective assessment will analyze the financial data benchmarks and review the pertinent market intelligence and then drill down into the details of the company's operations and the internal factors that impact them such as leadership and culture. The purpose is to build as complete a picture of a company's current state, warts and all, to prioritize recommendations and build the appropriate plans for improvement. In HRI's assessment, companies are scored on many hundreds of questions designed to penetrate into the corners of every department, program, and area. No process or function is omitted, no detail glossed over, and no stone left unturned. Management, sales and marketing, human resources, finance and administration, operations, engineering, materials, quality, and program management are all examined under an assessment microscope and rigorously scored on a rating scale based on best-in-class capabilities and future preparedness.

Our assessment process is purpose-designed to quickly and effectively diagnose the overall, in-depth health of your manufacturing company and, subsequently, arm you with the information needed for both narrow and broad-scale process and performance improvement. It moves rapidly from the gathering of raw data through the sorting and analysis of information, to the synthesis of knowledge needed to develop the understanding of why you are how you are and, ultimately, provide the motivation to change. But companies don't have to do the HRI assessment, the key is simply to engage in a process that assesses your business.

STEPPING OUTSIDE THE COMFORT ZONE

Many manufacturers want to keep their data private or discredit it because they're uncomfortable with the story it tells. This is basically a fear of not measuring up, and it's often explained away as an issue of trust. But here's the problem: How can you understand how you stack up in your market and against your competition? How can you aspire to be best-in-class if you don't know where the class is? Additionally, if you use the data to justify your status quo, the *condition* of your company, as opposed to using it to motivate improvement, then, in effect, you're claiming that the data is not reflective of your *perceived* reality—what you want to believe rather than what actually is. Leadership's willingness to explain away or doubt the data is almost always a cultural indicator of resistance to change.

ON-SITE AND HANDS-ON

The assessment process is also on-site and hands-on. Our on-site assessment is usually a two-to-three-day process, and we focus on what we see and feel, which is exactly what you should do, too. Look at the physical appearance of your facilities, the exterior details, storage yard, loading dock, shipping, etc. Are they clean or uncluttered? Organized? Is there attention to detail that would impress a first-time visitor such as a new customer or a job candidate? Does your site have curb appeal? Does it look well-kept? Is the grass mowed and the landscaping cared for, are the buildings painted and in good repair? If we drive around the back of your facility, will we see pallets neatly stacked, or will we notice a pile of crap in a corner of the employee lot? When you walk into the building, what's your gut feeling? Is there someone in reception to greet you? Do they make eye contact? Is there any kind of a *wow* factor? Is it a place where you'd say, "I want to work here"? Your data tells the story and provides an idea of how you are doing, but the initial impression is still necessary. It aids in placing your company in the correct quadrant of the performance matrix; then it's time to look for validation of key assumptions.

Walk the shop floor. Observe all your manufacturing processes—your lines, the equipment, the technology and automation you're employing, your labor and facility scheduling. Look at your plant, your front and back offices, your holding yard, your materials and inventory storage, and your loading docks with new eyes. As you watch the activity on your shop floor, the operations on your lines, the busyness of your offices, all the traffic through your physical plant, think about what you see and feel. Does the activity look chaotic or is there a sense of order and purposeful action?

For example, a general manager with thirty years of experience under his belt wouldn't acknowledge any of the challenges discovered during assessment. He simply didn't "see" them because when he was on the shop floor, he was usually moving through it with a singular purpose such as "I need to be on the other side of the plant for a meeting." His focus was task-oriented rather than full process or outcome-driven. All the various moving parts around him were separate elements or components. In effect, he couldn't see the forest for the trees. We suggested that he stand for thirty minutes at the center of the manufacturing floor with a notepad and do nothing but observe. He agreed reluctantly and at first seemed both physically and mentally uncomfortable, likely thinking "Why are they looking at me? What do they expect or want me to do?" while waiting for a shift worker to ask him what he was doing. But after ten minutes of discomfort, he started watching the activity on the floor and, a few minutes later, started recording what he saw. And after thirty minutes, he admitted that he'd never looked at the plant with an open mind without a specific purpose, and now he could see, or was beginning to see, what was really going on in the plant under "normal" operations.

When on the shop floor, most managers are purpose-driven, moving from point A to point B for a specific reason. Rarely do they pause long enough to observe the activity in the environment or note inefficiencies and areas for improvement. But to really assess the team/plant/enterprise they're leading or managing, people need to get to that place where they can "see" it with an open mind. In other words, they need to get out of their own way.

OPENING THEIR EYES

Managers and leadership need to look at "work" in a new way. Rather than "just get it done" motivation, they need to move to "get it done better." They should question everything they notice.

How are we picking up and putting down parts, and how many times a shift? How are we scheduling the facility? Where are the bottlenecks in the process and how can we open them up? Are our staff able to work as efficiently and effectively as we need them to? What can we do in this department, or on this line, or in this office to heighten productivity?

Plant and operational managers are too caught up in the minute-by-minute, day-to-day activities to see the big picture. Observing the business and manufacturing process organically is not an easy thing to do. No one really wants to take a hard, long look in that mirror.

We'll even send an owner or senior manager into a plant restroom or cafeteria and ask them, "Would you use it? Would you eat there?" In effect, we're asking them if they can put themselves in their workers' shoes. And every step of the way, we talk to your people, ask them what and how they're doing. What's their work environment like? What's their reality? Are they consulted, listened to, included? Do they feel respected, valued?

CULTURAL CHARACTERIZATION

It's important to assess your company's cultural characteristics and leadership style because, in truth, every culture is different. No one is right or wrong, but some are more effective than others. Leaders who are transparent and allow themselves to be vulnerable encourage employee buy-in, particularly in strategic motivation. Instead of mandating "Here's what I want you to do," they frame their business objectives inclusively. "Here's what I want us to do, here's why, and here's how I think we should do it... What do you think?" Sharing both reality and aspiration with employees is a sign of respect and endows the employee with ownership of the outcome. These leaders recognize that they don't have all the answers.

In contrast, tyrant leaders, although predictable and tolerated by their organization because they set clear expectations, demonstrate a lack of respect for their workforce and create an us/them environment in which the employee is a vehicle for success, but doesn't really benefit from it. In the latter case, we've found that attitudinal issues usually run throughout the organization. If you're fighting attitude, you're fighting a cultural mindset that's not conducive to change.

Another characteristic of leadership that is antithetical to improvement is complacency, a form of denial of the larger reality. If leadership does nothing about something that looks to be going downhill, that sense of decline will spread throughout the workforce to the shop floor. With company leaders and workforces in denial of the reality they face or are ignorant of the market conditions that are adversely impacting their business, the way the assessment insights are shared is critical to acceptance. One has to show how they're interpreting a company's data and then give them insights regarding what they've observed in the facility in a way that helps them understand why they need to change. We counsel our manufacturing clients to listen closely for denial and excuses, symptoms of complacency. If we hear excuses or statements from leadership or the workforce qualified with non-data-driven preambles such as "I feel..." or "I think...," then we know that the company's culture doesn't drive accountability.

Many companies work to foster open cultures, where free-flowing communication is encouraged and employees are willing to go to leadership and tell them "This is broken. We need to fix this." Unfortunately, there are also many companies where leadership is resistant to suggestion, and where employees are reluctant to express any negative observation or criticism lest they be labeled malcontent. In either case, the message is better received when supported by data.

Another cultural attribute to assess is communication—avenues of credibility, influence, authority, and ultimately trust. Look at how critical information flows in an organization and the nature and tone of intra-company communication. Does it come down from leadership as mandates and directives, or is it inclusive and participatory? At HRI, we believe that at the heart of company communication is influence. Good communicators are influencers who understand innately that their role is to coach and educate. The higher up the ladder you are in your organization, the more important this role becomes. Good communication, although often described as bidirectional (as in top-down or bottom-up), moves in all directions. We'll get into this more later on, but in a healthy company with an open culture, employees in every function and level of authority drive each other to higher performance when the mission, vision, values, long-range strategies, and short-term tactics of their company are clearly communicated and adopted. Your leadership can't only come from the top-down; it must come from all directions. This kind of cross-pollination hasn't always been welcomed in a manufacturing world in which a stay-in-your-lane mentality continues to prevail despite evolving workforce attitudes and preferences. That kind of narrow focus may improve the performance of a department but, at the end of the day, doesn't drive toward the enterprise-wide critical thinking that distinguishes Top Performing manufacturers.

REDEFINING INFLUENCE

Leaders, whether in the corner office or on the shop floor, are wise to look for teammates who make them think differently, challenge the norms, and push them outside their comfort zones. The people who are always aligned with your thinking only reinforce parameters that actually constrain your ability to transform or improve performance. If you build enough trust with those who think critically, then you can drive to better results quicker because you don't have to overcome preconceived barriers. This requires lateral influence—in effect influencing without authority despite the titles that say otherwise.

So how are influencers (lateral or otherwise) grown and identified? When promotions are used to bestow authority rather than responsibility, don't assume they'll be an effective instrument of change. Success of the whole is not achieved by improving only a portion of it, and there is a significant disconnect between authority and leadership. It's easy to find individuals eager to take authority and wield it, as in running a plant by inflexible rules. It's much harder to find someone who can influence operations without authority. Influencing without authority is a natural give and take. When people believe in you because you listen to them, they're much more likely to hear what you have to say. Those who can influence are the true leaders.

In the end, cultural shortcomings and issues can be major roadblocks to success, so our assessment tools are designed to catch them.

PRACTICE, EXPERIENCE, AND PUSHBACK

In addition to a company's performance and cultural data, one should also assess key experiential data, such as history and legacy, practice and experience—how long a company has behaved or acted a certain way. Look at how decisions are made, what inputs are used, and what the fallback "just doing business" habits are. For example, look closely at how forecasts are put together. Are numbers arbitrarily selected as target goals, or are they based on evaluation and review of what's been achieved in the past with certain customers over a specific period of time? And when were the peaks and valleys in performance and profitability? What do we know and what have we learned about those patterns? It stands to reason that an important part of assessing a company is also assessing what its customers are doing from a sales and marketing perspective as well. So, understanding how often customers are contacted and visited, whether data is being gathered on them, and how often they share their forecasts, is critical.

There's usually a certain amount of resistance when we begin assessing an organization. The misconception is that a consultant will simply "read your watch"—that is to say, look over your business and then charge you for something that's readily apparent to you or that you could have accomplished on your own. A good consultant *will* read your watch. They will look at the data you provide and consider your answers to their questions, but they'll see something that's not obvious to you, and they'll see room for improvement. That's what you should want and need a consultant to do. The revelation comes when a client is willing to accept feedback from a neutral party, i.e., someone who has no dog in the hunt or skin in the game. An enlightened client once said to us, "You read my watch ... but I was apparently in the wrong time zone."

To be clear, companies do not have to hire our team to do this assessment. The point is that companies should assess their business annually at a minimum and that can be done with a consultant, their board of advisors, a peer in the industry, or even a cross-functional team from within the company. But they have to be open-minded and willing to take a critical look at their business. Just like our annual physical, it's imperative to the health of our company.

That said, after we've completed a company's assessment, presented our findings and insights, and made our recommendation as to where to start a company's program of improvement, whether it's a business turnaround or fueling growth and sustainable performance, it's never our intent to be the smartest people in the room.

Our intent is to make the room smarter.

HAVE YOU CONSIDERED?

1. Have I assessed my business lately?

2. Am I open to feedback from my team and/or external experts?

3. Do I have access to all the data I need to make informed decisions about the business?

4. What type of culture do I have at my company? Is it an asset or a liability?

5. Am I willing to hear the good, the bad, and the opportunities to improve?

6. Can I admit that we are not perfect and open my mind to the art of the possible?

Gather the Data - financial, sales, operations

Commercial Strategy & Process

Technology Road Map

Leadership Sustainability

Vision/ Strategy

Talent Strategy

Analysis

Assess

Operational Experience

IMPROVEMENT PROCESS WHEEL

CHAPTER 2

OPERATIONAL EXCELLENCE AND THE IMPORTANCE OF BEING REAL

"When you're the lead dog on the sled team, the view is fine."

According to Dave Cecchin, president of Omega Tool Corp., a highly successful moldmaking company, they didn't feel a need for benchmarking since they knew they were #1 in their space. Additionally, their sector of the industry was fragmented, as he described it. Other tool & die manufacturers weren't open to sharing ideas or best practices, and Omega was no different.

"We felt we didn't need to compare our operations to anyone else."

We got to know Dave and Omega Tool when we were retained to audit all of Ford Motor Company's suppliers, including his company. At that time (years ago), benchmarking services weren't readily available for many second- and third-tier manufacturers, and few were attempting to benchmark themselves against their perceived competition. Dave and his company were not outliers in this regard. Furthermore, his resistance to benchmarking stemmed from how he defined "competition." To Dave and his management team at Omega, their competition was any tool & die enterprise that was trying to take their #1 status away from them. We defined competition differently. We believed (and still do) that Omega's competition was any company that competed for the same business, no matter how large or small the order. We told Dave that his competition was any company that wanted to be like Omega Tool. So, Dave agreed to benchmarking.

"At first, we were a little apprehensive about what benchmarking might reveal about our operations and performance," he admits. *"But now we're 100 percent committed to it—but not because we're trying to find out what our competition is up to. We don't worry about that."*

Instead, Omega Tool relies on benchmarking as a litmus test to track how they're performing compared to the entire tool & die sector of the industry. That is to say, how they stack up against the norms, the industry standards.

"We know that we're not first in every category. We may be great at three things and substandard at five others. So, we use benchmarking to evaluate our strengths and weaknesses, and to determine if we want or need to increase our performance in any of those areas and, if so, what are we going to do internally to make that happen."

In short, Dave uses benchmarking to identify areas of operational improvement. He is an edgy, hungry leader, and that is what defines good leadership.

—

When she first met us at a meeting of the Manufacturers Association for Plastics Processors (MAPP), Jenn Barlund, now president of Falcon Plastics, was working in the marketing department of the company. Not long afterward, we were retained to assess Falcon's three business units, which led to being hired on as business consultants.

"Falcon was stagnant. We had reliable, profitable business, but we weren't world-class. In reality, we were performing at sub-par levels. HRI came in and opened our eyes to how others in our space were doing versus our bottom-line results. They challenged us to ask ourselves what we could do to get our business to the next level," Jenn recalls.

"They motivated us and helped us realize that there's a difference between understanding your business, understanding where your peers are, and then understanding the potential you have. Now that we see where we are, we want to be world-class. We want to be top performers. They pushed us to open our eyes to the outside world, see what we could do

to improve and use the data to do it. Now we're continually driving to the next level and providing more for our customers. HRI encouraged us to be 'edgy' in their words. They want us to continue to grow and learn and be happy with our progress ... but never satisfied."

You simply can't do that without benchmarking.

First, let's explain what we mean by "operations." We don't mean operations on the shop or factory floor. That's far too limiting. When we talk about operations and operational excellence, we're talking about every single activity in a manufacturing company—on the shop floor, in the administrative offices, on the loading dock, in the supply yard, in maintenance and repair, on the trucks, in the HR office, in the company's social media and digital presence, in sales and marketing, in planning, etc. Every single function. Every single job description and at all times. Yes, it's broad, but it's broad for a good reason. Improvement in any one area can help the company. Improvement in several critical areas can arrest a downward slide and help a company find its footing in its competitive arena. Improvement across the enterprise that happens when everyone performs to a higher standard, to achieve higher performance, can and will take a company to the next level.

To do that, you have to establish goals and communicate them clearly ... to everyone. This is when everyone in the company becomes a partner in your success, from the corner office executive to the janitorial staff and even the team member who cuts the grass in the parking lot medians. Give them a goal and the reason why it's important and they'll become an "owner" in the success of the overall effort.

How you determine what those goals should be and how to drive toward them comes out of two key actions—thorough operational assessment and across-the-board benchmarking. Only then will you be able to identify the levers you need to pull to enact improvement, to transform operations, to take your company to the next level.

In a sense, assessment and benchmarking are two sides of the same coin. If you perform one without the other, you're only getting 50 percent of the picture. Assessment is a deep look into a company that defines *how it is* at a particular point in time. We equate it to a medical checkup on the health of your company. It needs to look at all the symptoms. A thorough assessment will accurately place your company in the correct quadrant of the performance matrix discussed in the last chapter and point to the areas that need work—that will likely have the most impact on the health and sustainability of your company.

Benchmarking, on the other hand, is a wider-angle view that describes *how your company performs* in its marketplace and points to opportunities for improvement. In addition to exhaustive review of a company's operational data, assessment deals with appearances, cultural hints, and impressions. Benchmarking is more about the performance numbers, analytical statistics, and industry standards viewed against the background of market conditions and local, national, and global economic factors. Assessment is *how you are*. Benchmarking is *how you stack up*.

The broad scope of operational assessment helps identify which levers to pull to begin the process of improvement, which particular areas to focus on and in what order. As we mentioned in the previous chapter, the assessment process should tell you where to begin, where to enter the improvement process and what needs to be addressed first—operations, labor, commercial process, technology road map, vision/strategy, or leadership and sustainability. When you conduct an assessment, you should look for ways to drive efficiency and improve the profitability of your operations by gauging how flexible or rigid you are in responding to market pressures. The more flexible you are, the more you can "bend" to drive efficiency while handling multiple business conditions such as supply chain and labor pressures or customer attrition, to name a few. Conversely, rigidity constrains your ability to handle more than one business condition at a time, and anything outside normal business operations becomes a problem. You should also be evaluating how you schedule jobs and allocate labor, checking your maintenance practices, looking at how you solve problems, what kind of processes you rely on, and whether you have a current, technologically enhanced set of operational best practices, or if you're relying on tribal knowledge and legacy processes.

Above all, your assessment, whether conducted on your own or with the assistance of a third party, needs to be thorough and all inclusive. You need to:

1. Define your gaps

2. Examine and benchmark your data (all your data)

3. Question everything

4. Get your people's input—What are they struggling with on a daily, weekly, or monthly basis?

5. Talk to your customers—How does your work/product affect their business? What, in their opinion, could you be doing better?

6. Be brutally honest with yourself, don't explain it away, whatever it is

7. Build short-term, mid-term, and long-term plans

8. Identify the things you don't need to pursue

9. GO SEE

10. GO ACT

11. Re-assess

GO SEE

The Go-See step that we discussed in the last chapter is what many leaders miss. It's a critical action in getting a true handle on the state of your manufacturing business. Leaders need to make their people go into their facilities and just observe the activity around them. Do they understand every action? What are their manufacturing instincts telling them? And leadership should participate too. What do they see when they stand still for thirty minutes or more and just watch what's happening on the shop floor, on that line, in that corner...

1. What "stops" them or gives them pause?

2. What's not "right?"

3. What's not efficient?

4. Where's the waste?

5. Is there purposeful movement/action or is it chaotic, disorganized, or inefficient?

6. Do operations seem rough and/or staggered or are they smooth?

If the leadership of a company doesn't have the time or can't bring themselves to take the time to walk out on the shop floor to Go See and communicate with your team about your observations, then you're likely not "in touch" with what's really happening in your company and you're not what we'd consider to be a "best in class" leader.

The Go-See process is critical to learning the true status of a company. But there's also a lot that can be learned about the nature of your company from the data and how it's received. For example, when companies aren't quick to provide the data we request before the on-site assessment, that can be an indication of which quadrant you may occupy. The poor-performing company may complain that they don't collect that level of data. Or they may not respond promptly to emails. Or they may not have the ability to pull data out of their own systems and processes, or know how to use it once it's gathered. Our response is always, "Don't create something. Just give us what you have." Our review of your data gives us an indication of where there may be operational issues, and helps us form the right questions for when we're with you on-site. Finally, how quickly you respond, and the nature of our pre-assessment dialog provides clues to the how and why of your operations.

When we're onsite, one of the first things we look for are the cultural "tells." These are the inadvertent behaviors or mannerisms that reveal true thoughts, intentions, and emotions or situations. These "tells" let us know how deep we'll need to dig to arrive at an accurate, full assessment. We'll do walk-throughs on all shifts to observe the mood of the second and third shift workers and see how they function in their environment. Do they know who we are and why we're there? If not, do they ask? All are indications of workforce investment in their company. Do they care about their work and the company or are they just going through the motions? We will also question workers on the shop floor. We'll ask, "What are you doing with this machine or with this process? What is the product? Where does it go?" They may not have all the answers, but if they're positive about their work and direct us to someone who can answer, that's an indication of a can-do culture. But if the first thing we hear from a worker is "I'm overworked and underpaid," then we have a good idea of what we're dealing with.

The point of the "Go See" is to understand the conditions (intangibles) and then bring the realities (tangibles) buried in the performance data to light.

Understanding company leadership interaction with their workforce is also a cultural clue. As you're walking through the facility, do you know the names of your people on the shop floor? Do the workers greet you by name? Most of the time, we can get a sense of the culture right out of the gate. If workers question our presence, then they may not have been informed why we're there. No one told them that management was bringing in consultants to assess and benchmark their operations. Naturally, this indicates broken communications and a lack of transparency. We've found that a tone of underlying disinterest or resentment usually comes from management's style of leadership.

Assessing a company in a short period of time can be challenging and in most on-site assessments, the second day is easier than the first. We have learned that by approaching an assessment with respect for the people and the company, cooperation is much easier to obtain, and we have a smoother path to validating what the data is telling us with the psychological indicators. We don't go in assuming that we'll find problems. An assessment gives us a baseline picture of the general health of a company. Consider again the analogy from chapter 1: When you go to a primary care physician for a checkup, your vital statistics such as height, weight, blood pressure, pulse—the data—are recorded in your chart. The office visit or consultation during which the doctor examines you is when the doctor puts your data together with physical observation to get a more complete picture.

To sum up, assessments include a review of a company's financial and operational performance data (KPIs), and strategic plans for labor, sales, and marketing, and a "Go See" on-site walk-through to get a sense of the intangibles (those attributes or conditions that can't be quantified) that impact the health of a company. We also evaluate a company's vision/mission/values for coherence and allegiance. Are they meaningful and embraced by the workforce, or are they just words on a poster in the breakroom or on a page on the website under "About Us."

Once performance baselines are understood and validated, then you can begin to identify opportunities for operational improvement. Take physical and transactional waste, for example. Physical waste is visible. Transactional work, such as project management, engineering, sales, and marketing, are process-driven and must be thoroughly grasped in order to evaluate efficacy and efficiency. The questions we ask reflect this need for understanding: How do you project manage? How do you engineer? How do you sell? How do you know it's efficient? Only then can you Go Act.

INPUT, OUTPUT, THROUGHPUT

At the heart of any discussion about achieving operational excellence is throughput. But what exactly is it? Also known as flow rate, throughput is a measure of a business's inputs and outputs within a manufacturing or production process. As such, understanding a company's throughput is a key part of operational improvement as it indicates the overall efficiency and flexibility of the entire organization and all of its processes and activities. Throughput is a company measurement, not just a direct labor measurement.

Consider supply and demand. During supply chain disruptions, manufacturers strive for a one-to-one match of demand and supply.

If you, as a manufacturer, focus on throughput with the intention of matching supply and demand to achieve the flexibility with which your company can respond quickly to any customer request, market condition, or unexpected occurrence (e.g., pandemics, natural disasters, etc.), what you're really doing is focusing on what you can control. The market controls prices, raw material costs, and packaging costs. So other than controlling internal scrap rates, as manufacturing managers all you truly control is overall process and labor efficiency.

For example, if a manufacturer builds a product of poor quality, the product has to be redone and more waste is created. With focused operational improvement, process, equipment, and training can be put in place to standardize work to a specific quality level, thus minimizing waste. When a company says that they're measuring defects or some kind of constraint in their manufacturing process (like a bottleneck), it invariably ties back to throughput issues such as a lack of good quality control standards or a flawed process. Another example is if a company is forced to raise wages in order to retain an able workforce, then that outlay will need to be offset through operational improvement in order to do more with the same human resource—and not just the direct labor on the shop floor, *all* labor.

If your throughput increases, then you're performing better collectively. But remember—while throughout is critical operationally, it cannot take priority over safety and quality.

LOW-HANGING FRUIT

Two of the most obvious areas ripe for improvement in most manufacturing operations are inventory management and scheduling. Are you looking at how often you're turning inventory, and how many days to turn cash from a purchase order into a product received? The truth is that inventory can be used to mask other challenges. Some companies build inventory to protect against the inability to make a shipment, or to make up for poor quality, or to accommodate unscheduled downtime due to maintenance. Inventory is built up as a hedge against a host of things that can go wrong. So, taking an unblinking look at inventory levels can point to other operational issues. Since many manufacturers are contractually required to hold inventory for certain customers, we also look how much inventory is controlled by the manufacturer versus how much is contractually mandated by the customer.

On the other hand, extremely low inventories can signal through-put issues as well. For example, if you sell 1,000 units of product per month to a steady, repeat customer, you might consider running a lot size of 12,000 and put 11,000 in inventory. Then you can fill the customer's monthly order out of inventory while freeing up that production equipment or assembly line to drive additional throughput. The cost of carrying that extra inventory is lower than the costs of the additional setups. However, it's critical to get the contract terms and conditions of this arrangement right and agreed upon from the outset.

We often use a swim lane analogy when we address scheduling issues. In swim meets, the fastest competitors are usually grouped in the center lanes of the pool with the slower swimmers assigned the outer lanes. In this way, the fast swimmers (your high machine hour parts) are not impacted by turbulence or the "noise" caused by slower swimmers (your low volume-high mix parts). Using this concept, it makes more sense to run high-capacity tools together to improve throughput rather than allowing lower volume, high change over frequency parts to decrease it. The improvement challenge in this analogy would be the need to modify the operational standards on the low-volume tools, i.e., the slow swimmers. Consider putting the low volume tools in their own pool so to speak and create a plant within a plant driving the outer lanes to increased performance. In other words, what will it take to bring the high mix lines to a different and more appropriate operating standard?

Inventory management and scheduling may be the easy targets for improvement, the low-hanging fruit, but driving operational excellence is much more than fixing a process, revving up a line, or thinking tactically about filling orders. The scope of operational improvement, and its potential, reaches across all business operations. If you dive into process and technology improvement, it's not just about the machines and people on the shop floor. It's about everything that has to happen to take a product from idea to delivery, including how we design the product and how efficient is that process. When we have to make a new part, do we consult design libraries to look at past projects of a similar type and size to leverage that captive knowledge—or do we take a blank sheet of paper and design it from scratch? Do we build on what we know and then apply new technology to manufacture it more efficiently? Do we use everything we can to make the operation better? Are we leveraging all the assets we have on hand for the benefit of the total business and most importantly, are we demanding zero defect mentality in our upstream engineering functions?

It's easy to talk about operational improvement as only the hands-on manufacturing part of the business. But it's far more encompassing than that. It's all the capabilities in all the departments. It's leveraging the engineering and business management software, the product design, the machine sizes, and types. It's looking critically at how you sell your capabilities and outputs, with certain commitments or design requirements. How do we hand off from sales to program management to engineering? What's the business process flow? What aspects or functions—sales, general administration, engineering, design, etc.—are impacting general efficiency? Did we quote and cost that order correctly?

THE POWER OF CRITICAL THINKING

At an industry icebreaker, a customer hands you a cocktail napkin with a product sketched on it. It's the idea he needs realized, and he wants you to fulfill it. How do you design it efficiently? If you're thinking critically, you might leverage past designs of similar products you've produced and discover that if you tweak this design or that feature, you could make it more cost-effectively and efficiently. If you're not thinking critically, you might look at that cocktail napkin and see only a manufacturing problem you may not be able to solve, thus running the chance of losing that customer. Thinking critically about every portion of your business, knowing your market, understanding your customer, every phase of manufacturing, how every action either helps or hinders the process flow upstream or downstream is vital to operational excellence. Critical thinking leads to efficient problem-solving.

We observed critical thinking in action at a coffee bar in an airport terminal that we move through on a regular basis. This coffee bar always seemed to have long lines. One day the lines were much shorter and when asked, the manager explained that he had matched the TSA schedules with his labor allocation. When TSA opened another lane at the security checkpoint, he added another barista to handle the increased traffic on the concourse between security and the gate lounges. No one told them to do it. Observing a bottleneck in the process and, thinking critically, provided a solution. A simple example, to be sure, but it underscores the point. Critical thinking begets efficiency.

Bringing the discussion back to manufacturing, the worker on the shop floor knows best about what they're dealing with in regard to process, workflow, bottlenecks, etc., but they're not necessarily encouraged to think critically about the throughput issues they see on a daily basis. They're not often engaged in creative discussions about bettering the company they work for. If the company culture doesn't provide opportunities for people to "attack" safety, quality, and throughput issues, it should. This ties back to what we see and sense when we do on-site assessments, when we watch for how people behave on the shop floor and the tone of communication between leaders and workers. The more open the culture, the more likely the company is open to modification and willing to change for the better. When you engage and empower people to think about how to do their jobs better, you're encouraging critical thinking. And when the legacy generation passes down their hard-earned knowledge to the upcoming generation of leaders, something we'll discuss more in later chapters, you have the power of not only changing your organization—but the industry.

DATA DOESN'T LIE

We've explained that improving throughput is at the heart of operational excellence. But how do you evaluate it beyond the obvious areas of inventory management and scheduling? You have to go back to the data and examine it with a clear eye. Identify the top-performing areas and the outliers and pick the top ten areas and the bottom ten. Then ask what can be learned from the top performers that can apply to those areas in need of improvement. If the data doesn't tell a clear story, then you "cut" the data again and again until you find something you can work on. It's not refining the data until it tells you what you want to hear. It's digging deeper until you find the truth of the situation—the real condition.

In most cases, we use Process Quantity Analysis (PQ) when we "cut" the data, as in what process and what quantity are we looking at? For example, we might look at:

1. Sales dollars by customer

2. Sales dollars by part

3. Run hours by machine

4. Throughput of second (or third) shift

5. Actual cost versus standard cost

6. Quoted cost versus actual cost

An example is maintenance, which is often assumed to be outside a company's control. The real story is in the data—in the work orders and machine run times, preventive and predictive maintenance schedules, and the frequency of unplanned/catastrophic failure. The better job maintenance does in breaking out the demand on a maintenance person, a skilled technician, regarding what they can and can't control, will drive efficiency. For instance, maintenance can control when they do preventive maintenance on a machine based on level of use or length of operation, while aspiring to predictive maintenance and negating the need to run to failure. Maintenance should be spending their time maintaining, not repairing. Use the available data to not just plan maintenance but start predicting potential failures, which minimizes the unplanned downtime and frequent run to failure.

The best leaders and top performers combine analytical capability with data-mining capability to figure out how to move along the path toward operational excellence. Again, that younger generation of new engineers or team members can help us get there. They can analyze the data and then we can teach them how to make the improvements. This is where an honest assessment benchmarking and critical thinking all come together. This is where you envision the art of *your* possible and determine how far you can push your throughput, how much you can improve your income statement and balance sheet, what you can do to overcome your labor shortage or supply chain disruption.

Like all continuous improvement methodologies, it's a circular process. You use assessment and benchmarks to determine the need for improvement in a given area. You work the problem, devise a plan, and enact it. You analyze the data and measure your improvement against the benchmark. Then you re-benchmark the performance and start again.

It's important to keep in mind that this process is a journey, not an activity. The assessment points you where to start, but you shouldn't overengineer process improvement. This can't be a lean initiative or a kaizen event. It's basic blocking and tackling; it's using the data in your company to work on just one or two things to help you lower the water level. You have to be data-driven, and you have to provide your experienced team members with the resources to teach and leave a legacy with the new workforce generation that wants to be efficient and make an impact. Turn your teams loose on whatever the bottleneck is and let them try things. You have to *act*. You have to get up to the plate and just swing—even if you only get a successful hit three out of ten tries, you're still batting 300. Operational improvement is a culture to drive to be the best—it's the further development of your tool kit so you can transition experience to the next-generation employees, who will create sustainable operational excellence.

We could not possibly cover every operational topic that could come up in a manufacturing company in this chapter or even this book, but we can—and will—provide you with the foundational resources you need to ensure that your team understands how to utilize critical thinking to tackle a problem. Often times we overcomplicate things when we just need to stop and analyze the situation. When you talk about improvement and driving toward operational excellence, it's important to be honest with everyone about what the data says and where the opportunities lie. We maintain that improvement shouldn't mean being as good as the guy next to you or your nearest competitor. It's about performing better than you were last week or last month or yesterday—every day. It's about using your own benchmarks to figure out what you can improve on. At some point, you'll need to ask, "What's my competition doing that I'm not? What are the best and brightest in my space doing in this market climate?"

And then you do it better.

HAVE YOU CONSIDERED?

1. When a potential customer or a job candidate visits your facility, what is their first impression? What do you want it to be?

2. Are you benchmarking your operational performance internally (against past performance) or externally (against your competition)? Why?

3. How would your workforce react to an on-site assessment?

4. When you walk through your facility, what's your first impression? What do you notice first, what's working or what could be improved?

5. Do you know your workers and how they feel about their work and the company? Have you ever asked them?

6. What is your "low-hanging fruit" of improvement opportunities?

7. What operational functions outside your shop floor—sales, general administration, engineering, design, etc.—do you think impact throughput, and how?

8. Do you encourage your workers to think critically?

9. How would you characterize the "culture" of your company?

10. If your operational data doesn't paint the picture you'd like to see, what's your first instinct—to disprove the data or dig deeper to find out why?

11. Do you share operational and financial data with your entire workforce, or just with department heads? Does your workforce know how they're performing?

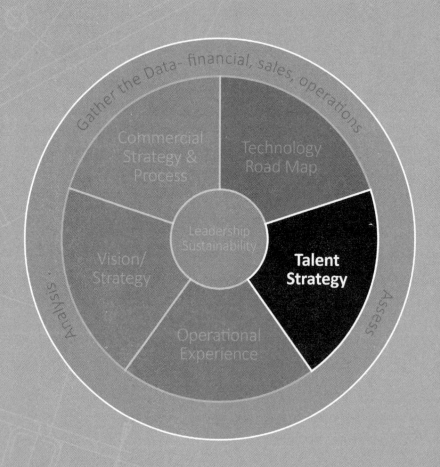

Gather the Data- financial, sales, operations

Analysis

Assess

Commercial Strategy & Process

Technology Road Map

Leadership Sustainability

Vision/ Strategy

Talent Strategy

Operational Experience

IMPROVEMENT PROCESS WHEEL

CHAPTER 3

THE LABOR CHALLENGE THAT JUST WON'T GO AWAY

"We can't replace a skilled worker with twenty years of experience with another one with the same level of experience. That person simply isn't out there. All the good ones are already working and to recruit them away is going to be expensive. The only available pool of talent are Millennials and Gen Zs who work and think and communicate differently. If we don't learn to accommodate their differences, take their strengths, and build on them then, as an industry, we are destined to extinction."

—Dave Cecchin, Omega Tool

—

"Anytime you're trying to get a group of people to change and adapt to new conditions and new ways of working, it's going to be challenging and you're going to get push-back. Especially during a time when technology is moving a mile a minute. We have employees who have been with us for forty-plus years, and for them change is hard. On the other hand, the younger generation is comfortable with change and technology because it's been a constant in their lives. But the fact remains, tooling and skilled manufacturing tradespeople are really hard to find, and we need to retain as much of that expertise as we can. We don't want to lose that tribal knowledge. So, yes, there's a disconnect between the older and the younger workers, but each group has their strengths. The younger generation brings a lot of fresh ideas to the table and meets technological challenges in new ways and without fear. So when a young employee says 'I've got an idea, let's try this out' and an older employee responds with 'Tried that, it doesn't work,' I'll pull them aside and tell them 'Hey, it's a different time, If we're doing something the same way we always have despite the availability of new processes or tech-nology, maybe there's a better way to do it now.' It takes a lot of coaching but, eventually, you can move them away from dismissing ideas or suggestions because they're not familiar with the technology that can enable improvement."

—Jenn Barlund, Falcon Plastics

alk to almost any small-to-medium manufacturer these days and they'll tell you that the biggest challenge they face on a daily basis is the shortage of labor. According to our research, Baby Boomers and Gen Xers make up more than half—55 percent—of the workforce on average. Of the remaining 45 percent, Millennials account for approximately 33 percent.

Average 2023 Workforce Demographics

- Boomers (>56 years): 21%

- Gen Xers (41–56 years): 37%

- Millennials (25–40 years): 32%

- Gen Zs (<25 years): 11%

Workforce by Generation

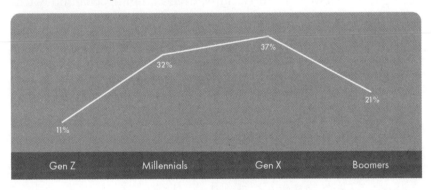

Figure 3.1

That means that the youngest talent in the national pool, the Gen Zs, make up only 12 percent of our current manufacturing workforce. Based on this, more than half of the workforce is less than twenty-five years from retirement age, and half of that is looking at leaving the industry in less than a decade or so. In fact, the "Great Retirement" has already begun. With the impact of Covid and the involuntary furloughing of shift workers as shops were forced to shut down, many of those workers are deciding not to come back. In a sense, the Great Retirement could also be called the Great Reckoning in which workers are being forced to look at how they want to spend their remaining years.

The problem most manufacturers face today is how to attract new, qualified talent to backfill the outflow of skilled labor. The facts are pretty bleak. Fewer and fewer young workers are willing to consider manufacturing as a career choice because they've been directed by parents and school counsellors away from it. Manufacturing is still viewed as a dirty, noisy, dead-end blue-collar job when, in reality, today's manufacturing requires clean facilities and highly skilled, technology-savvy workers and informed, astute management with an expanded world picture.

Then there's the generation gap found in almost every manufacturing facility in North America today. Old school versus young upstarts. Or analog versus digital. Baby Boomers and Gen Xers simply think, process, and execute differently than Millennials and Gen Zs. Boomers and Gen Xers are forced to adapt to the introduction of technology, new processes, and new paths to improvement that moves them further and further from how they used to do things. Change is not only scary, it's uncomfortable. Millennials and Gen Zs grew up with it. To them, continuous change is simply a fact of life. They look forward to the next best thing and won't hesitate to try it. That may seem to be an oversimplification of generational differences, but that's at the heart of it. Call it a culture clash or a paradigm shift, but it's undeniable that older workers and younger workers simply don't perform the same way. They have different concepts of what the workplace, be it shop floor or engineering, loading dock or administrative offices, should look like and feel like and, most importantly, require of employees. Obviously, it's not the same old manufacturing plant anymore. And it's not the same old workers, either. To a Boomer or a Gen Xer and to many manufacturing business owners (according to our research, 74 percent of whom are older than age sixty), that's a bitter pill to swallow.

LABOR PLANNING ON A SHIFTING PLAYING FIELD

Allocation of labor and scheduling when you're facing continuing labor shortages is particularly difficult. You're making decisions that are strictly short-term. "What do I need for today? How many people do I need in the plant today? Where do they need to be? What do they need to be doing?" Forecasting and demand planning during market volatility and supply chain disruptions is hard enough. Then add the challenge of putting the right people in the right place at the right time when you have absolutely no one to spare. When demand is difficult to understand for whatever reason, responsiveness to labor planning needs to be immediate. Agility and flexibility are critical. You need to change or pivot depending on customer, market, facility, economic conditions, etc. You have to be able to plan labor almost on a shift-by-shift basis. "Do I have enough people to run everything we need to run today?" If the answer is "no," then you have to make a change that doesn't necessarily mean shutting down a machine or a line. It may mean taking someone from one job function (maintenance, for example) and putting them in another function (such as running a machine) which, in turn, requires cross-training. But how can you do that if you can't find enough talent willing to come to work for you?

Recently we became aware of a $250 million manufacturer that was onboarding twenty to thirty next-gen people (temps and new hires) a week, but only a couple would actually stick around because the onboarding wasn't comprehensive, the plant was dirty and hot, and safety wasn't emphasized. These new hires looked around and said to themselves, "Even at $18/hour this isn't worth it. I can find a better job where they care about me, and the work environment is cleaner and safer." Furthermore, the people tasked with training them indicated that they wouldn't invest much time in the process as they figured they wouldn't be around for the long term. The trainers' frustration coupled with the less-than-satisfactory work environment and culture served to drive new hires away—effectively wasting up to $18K in hiring and training costs per new hire.

Evaluating the current labor/talent status quo is an important part of understanding your business. In many cases, it's the most important. After all, the people run your business, not the machines. And this is true even in highly automated operations. The questions to ask and the areas to probe include:

1. What does your company culture look like?

2. How robust is your onboarding plan?

3. Does senior management take time to talk to and get to know new employees?

4. How often do we follow up with new people to get feedback?

5. How clean and attractive is our plant? Would you let your kids work there? Would you work out there? Would you use the restroom?

6. How do you attract the talent and skilled labor you need?

7. How do you retain them?

8. What are you doing to empower and engage your current employees and new hires in the processes of your company?

9. Are your workers just bodies running machines, making parts, and packaging them? Or are they treated as part of the team, solving problems and engaging in activities that make the business better?

10. Do we market our business utilizing our marketing resources so that when the next gen goes on our website or to Glassdoor they see a place they want to work?

For many business leaders, the answers to these questions, or lack thereof, shine a light on many of the labor shortage issues that have gone unnoticed and unaddressed, while underscoring the need for nothing less than a paradigm shift in how they recruit and who is recruited.

THE NEW LANDSCAPE

Fact: It's a changing world. You can't recruit and retain the way we did one or two or three decades ago. For perspective, Indeed and Glassdoor were launched in 2004 and 2008, respectively. But Glassdoor didn't add actual job listings until 2010, and company ratings (pay, pay by race, ethnicity, gender identity, etc.) weren't added until 2021. In the "old days," which weren't really all that long ago, the effects of institutionalized and cultural racism and sexism meant that women and BIPOC (Black, indigenous, and people of color) individuals were, if not intentionally excluded from these opportunities, at the very least overlooked in favor of candidates who resembled an organization's status quo, the people who built the companies—middle-aged, white men—and that's who was looked for when it came time to hire. There was a belief that certain genders, races, and even ethnicities couldn't or wouldn't operate effectively in manufacturing. It was a white man's world, to put it bluntly, a male-dominated arena, particularly on the shop floor and in the skilled trades.

Today, diversity, equity, and inclusion (DEI) policies dictate that more women and minorities be included in manufacturing workforces. The problem is manufacturing, as a sector, has been a historically male dominated space—and in leadership, a traditionally white male domain. Did you know that since as early as 1987 there have nationally been more women college graduates than men in America? And yet for decades women have made up only about 31 percent of US leadership roles.[2] Women are entering the workforce in droves, but they aren't being represented in leadership. When a candidate who is a woman, African American, Latino, or a mixed race looks around during an interview and fails to see anyone like themselves on the shop floor, in the management offices, or in leadership roles, chances are good that they're not likely to accept employment. This is particularly true for next-gen workers. They're looking for diverse environments where skills are valued.

---

2   Chris Gilligan, "States With the Highest Percentage of Female Top Executives," US News, March 6, 2023, https://www.usnews.com/news/best-states/articles/2023-03-06/states-with-the-highest-percentage-of-women-in-business-leadership-roles.

It was also a manual environment. That is to say, you utilized an individual's skillset rather than applied technology or a group capability to accomplish tasks. For example, Joe was good at X and Bob was experienced in Y, but neither worker was ever challenged to learn or train outside their "lane." This was the rule rather than the exception. The result was that companies developed a lot of tribal knowledge where people were knowledgeable about small patches of responsibility with preconceived start and stop points. They weren't process-driven or aware of the upstream or downstream impacts of their efforts. It follows then that when skills and functions are so stove-piped or segregated, enterprise-wide operational improvement, which is dependent on transparency, cross-training, open communication, and shared objectives, would be a very tough nut to crack.

BATTLING SHORTAGES AND BUSTING STEREOTYPES

A key question is what level of higher education is required for manufacturing? Fifty or sixty years ago when manufacturing jobs helped develop the American middle class, you didn't need an advanced degree to get in on a ground-floor opportunity and then move up the ladder. Manufacturing jobs were good to have, and many workers stayed with their employers for decades, some for their entire careers. But it was hard work and many parents of that generation (Boomers) wanted more for their children. "Don't do what I did," they'd say, meaning don't go to work right out of high school. Boomers and Gen Xers encouraged their kids to go to college, get a higher degree, and join the professional class. School districts (and society) rewarded counsellors for the percentage of students that continue on to four-year universities, thus pushing more students away from the skilled trades. That was their definition of success. Unfortunately, it led away from the industrial/manufacturing sector.

But in the last four or five decades, our world has become a more complicated, competitive, and crowded place. Universities offer far more diverse curricula along with highly specialized career options to a larger and continually growing demographic. It's likely that it's only a matter of time before there will be more Millennials and Gen Zs entering the job market than Boomers and Gen Xers departing it. The problem is manufacturing just isn't all that attractive to them because they've been warned off, and the result is a twenty-plus year gap in the workforce.

So where do you go to attract new hires? You go to where the kids are—and you talk to their teachers and counselors (who in turn can be taught to talk to parents). Manufacturers won't be able to locate, train, and retain the talent needed unless they go into the schools and universities and underwrite programs, sponsor internships, and apprenticeships to snag the attention of the next-generation workforce.

—

Falcon Plastics with facilities in South Dakota, Tennessee, and Oregon, faces a kind of double whammy in regard to labor. In addition to a general scarcity of skilled labor market-wide, three out of Falcon's four plants are located in small communities with equally small talent pools.

"We simply don't have access to the skilled labor we need," explains Jenn Barlund, president. "With most of our new hires, we're forced to do a lot of training to get them into the roles we need to fill. During Covid, we had to furlough some of our people. But in 2021 demand exploded and meeting it

with a smaller workforce has been, and continues to be, a challenge. So, we asked ourselves, how can we control in this situation? We decided that we could control automation. We could focus on how to run presses without people—integrating automation where it made business sense, not to replace people but to do more with what we have. What we're doing is trying to minimize burnout among our team members by supplementing our manufacturing capability and capacity with automation."

But that's not the only labor initiative at Falcon. Jenn explains that Falcon has robust apprenticeship programs in most of their facilities. They worked with state authorities to lower the apprenticeship eligibility age from eighteen to sixteen years so kids could come in after school to learn various trades and skills.

"Every year, we send Falcon representatives around to the local high schools to do program interviews. The kids pick their top choices in apprentice programs and if they pick us as their first or second choice, then we hire as many students as we have slots to fill. We give them flexible work schedules (they still have homework to do) and move them through different positions, functions, and departments so they get a well-rounded idea of what manufacturing is about and what it can offer them career-wise. Some kids come back year-after-year, and then opt to stay and work for us right out of high school."

Overall, Jenn is impressed with this generation's work ethic. "These kids have grown up in a weird, turbulent time marked *by uncertainty. They work smarter, not harder, and will go straight to the solution. They're more interested in efficiency."*

She has also observed that, in general, Millennials and Gen Zs are early adopters and adapters.

"They're willing to change because they grew up with a lot of it. They pivot to new ideas and new technology quickly, whereas our older workers need to sit back, see how it works, and how it goes before getting on board."

—

Another challenge that many manufacturers face when trying to attract new talent is overcoming the negative impression of the manufacturing work environment. What is the condition of your facility? Is it clean? Dirty? Is it clearly organized or does the shop floor look confusing? Is it well-lit? Are the breakrooms and bathrooms clean? What does the front office look like? What is the ratio of men's restrooms to women's restrooms? (It's a sad fact that many manufacturers don't even have women's facilities on the shop floor.) We want manufacturers to ask themselves, "Would my friends or my kids want to work here?" if the answer is "No," then the next question should be "What needs to change?"

Perhaps the most dangerous stereotype commonly found in manufacturing is the belief that Millennials and Gen Zs don't want to work and won't apply themselves when asked to. It's an unassailable fact that they're digital natives and are more intelligent in ways that Boomers weren't taught to be. They're looking for work where they can make an impact, where their effort has meaning, makes a difference, where their contribution matters. They're not interested in just being another body on the line or on the floor or in the office. But in many cases, that's exactly how the older generations regard them, suspicious of the new ideas and processes they bring to the party. So, the younger generations leave because they're treated poorly, and then the old hands call them quitters.

IT'S TIME FOR CHANGE

In the manufacturing world of our fathers and grandfathers, independent thought wasn't encouraged. The operational line was "This is how it's done, so why would we do it any other way?" Older generation workers are quick to dismiss younger workers' ideas, writing them off as having little "real world" experience and that they need to "pay their dues" or earn their right to have a voice. The fact is the older workforce thinks and processes information one way because that's how they were taught. The younger, technologically confident workers think, process information, and view the world quite differently. The notion that the younger generation needs to adapt to and embrace old ways is just flat-out misguided. It's counterproductive. The aging workforce needs to adapt, retrain, and embrace the new. It's no longer something to debate—it's necessary. And, after all, it's more efficient. It's less wasteful in terms of time and cost. And because not too far down the road this younger generation will not only own the purchasing power, they'll also be running the show, especially with trends showing that workers are retiring earlier. Like it or not, they're the future.

"They don't know what they're doing. They haven't experienced it."

"They don't know how to leverage technology and aren't interested in learning."

The solution to this generational disconnect is to change the nature of the narrative away from old versus new, us versus them, proven versus unproven. Where an older worker might state, "Here's how I learned to do my job," the younger worker, armed with technology not available ten, five, or even two years ago, could be encouraged by asking instead "How would you approach learning this job?" If it's a two-way discussion that melds experience/expertise with new thinking, new tools, and new methods, then it's a win-win where both "sides" are contributing the best of what they have to offer. More often than not, the older generation says, "This is what we should do," while the younger generation protests "But we have unique skillsets and tools to solve problems and the technology to do it faster." Manufacturing needs to harness that desire to be efficient, use the data/information at hand, and marry it to the old-school expertise of the older generation—instead of either group writing the other off.

In a perfect world, both older and younger generations would accept and appreciate their differences and recognize the innate value of both. People are a business's #1 asset. That truth has never changed and, during a time when fewer people are attracted to manufacturing, we need to be even more effective in keeping the people we have, utilizing every aspect of their skillset to do better, while being more accepting of the new, advanced skillsets of the up-and-comers and their life priorities and values.

But for now, immovable object meets irresistible force, and manufacturers have to figure out how to bring the two together for the benefit of the whole. And to achieve this, there needs to be cultural change by leadership. Leaders need to be intentional in getting your younger and older workforce to align. What the best have done to attract women and people diversity is to hire multiple women or diverse employees at a time so they can build a sense of camaraderie. This allows them to acclimate more easily and provide a higher chance of retention. This kind of change will not just happen, and it certainly won't happen overnight. And it might mean you need to get rid of some people who won't change.

—

When Omega Tool in Ontario, Canada, realized they were facing a serious shortage in skilled labor, they actively recruited candidates from high schools and technical colleges, even first-year university students. They sat this group of new hires down and asked them "How do we prepare the company for the future?"

"Why are you asking us?" they wanted to know.

"Because you are our future," was the reply. "We're already on the retirement path."

So, they started picking their brains, asking these young workers, "What can we do to make this place, your work experience better? What will it take to make your friends want to come work here?"

Among other suggestions, one of the counter-questions that came back was, "If I'm not working on the shop floor all the time, why can't I work offsite? Why do I need to be on site for an eight-hour shift?"

In mid-2019, Omega Tool assigned a committee to look into the possibility and logistics of remote work, and to evaluate the communications and remote meeting platforms available at the time. Eventually committing to one, they started a six-month test of remote work practices. Then in March 2020, Covid hit and the Canadian government recommended that workforces be sent home to minimize the spread of infection.

Dave Cecchin, president, told his people, "We've already got people working from home. The trial period is over. Send everyone home except for the folks working on the factory floor." His reasoning was that with thousands of square feet of space, sixty-foot ceilings and air cleaners, no one needed to be closer than six feet from each other. The company staggered their shifts, changed lunch break patterns, kept everyone on-site separated, provided all necessary PPE, and everyone else worked from home. In the end, Omega Tool did not miss a single day of production throughout the pandemic. In fact, within two weeks of the initial shut-down mandate, as a preferred supplier to GM, they developed tooling for making ventilator parts and went on to be a key part of the effort that increased ventilator production 300 percent.

Had they not taken the time to listen to the youngest members of their workforce and be willing to act on their suggestions, Omega Tool wouldn't have been able to react to the global pandemic as quickly and effectively.

DIVERSITY AS OPPORTUNITY

Once when we were facilitating a strategic planning session for a client, we were involved in a big discussion about labor and the future of this particular client's labor strategy and what needed to happen to bring more diversity into their organization. Their head of human resources said, "I'm going to do this, and it'll be different." We looked around the table and saw eleven middle-aged white men and a single woman. So, we asked, "How are you planning to do that?" and the HR guy replied, "Well, you know, we just will." "With all due respect," we replied, "You've been in this role for twenty years and you haven't done it yet. What's going to change?"

—

Lack of DEI in the workplace is a genuine turn-off for the Millennial and Gen Z job candidate. If you don't see yourself in the workforce, it's not likely you'll feel comfortable working in it. Many manufacturers simply don't understand that. A manufacturing shop floor—not to mention the front office and leadership—should look like its surrounding community, diverse in gender, race, and ethnicities. Often when we assess a company we'll ask, "Do you have women in your facility?" and more often than not, the reply will be "Yes, they're in marketing" or "They're in human resources." But we've noticed that those of our clients who have women working in their plants, on the shop floor, get better results not because they are women but because diversity has proven to be more efficient. Problems are solved more quickly because women have different ways of approaching and analyzing problems. Twenty years ago, 98 percent of the manufacturing population were white men. Today, women make up approximately 30 percent of the manufacturing population, still way out of balance when you consider that women account for more than 47 percent of the nation's workforce.[3] The current manufacturing demographic doesn't reflect the national reality.

We study culture, diversity, and workforce among manufacturers and found that the industry is less diverse than the overall US workforce. Women make up only 19 percent of management in tooling and 30 percent of management in production; and only 13 percent of the overall total workforce in tooling and 21 percent in production. Black, indigenous, and people of color (BIPOC) of either gender have little to no presence in management in neither tooling nor production, and account for only 5 percent in tooling and 33 percent in production. In 2022, the manufacturing boardroom was still white men, middle-aged or older.

---

3   Earlene K.P. Dowell, "Manufacturing Opens More Doors to Women," US Census Bureau, October 3, 2022, https://www.census.gov/library/stories/2022/10/more-women-in-manufacturing-jobs.html.

Manufacturing is Less Diverse Than US Workforce

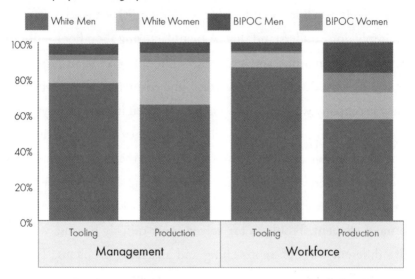

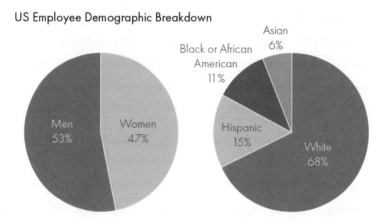

Figure 3.2

Diversity is also about generational mix as well. When we look at the manufacturing universe compared to the overall US workforce, Baby Boomers comprise 20 percent of the workforce, aligning with the national average. Gen Xers comprise almost 40 percent of the workforce, running fifteen points above the national average. Millennials make up at 32 percent of the workforce 14 points below the national average, and Gen Zs account for about 10 percent, as compared to the national average of 10 percent. This means that the oldest group of the workforce is headed off into retirement at a rate that is twice that of the incoming new talent. Unless manufacturers make a concerted effort to attract young talent and assimilate their strengths, the generation gap is likely to get even more pronounced.

Manufacturing is Short on Gen Z

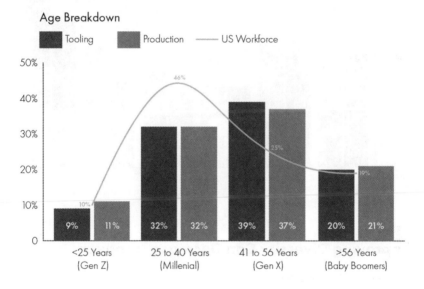

Figure 3.3

Most owners of manufacturing companies, as per our research, are sixty or older, which makes long-range planning, particularly leadership succession, critical. Lack of succession planning leads to disruption on multiple levels. The older the leadership in a company, the less likely they are to trust the future of the company to the younger generation. Conversely, the younger the leadership, the more likely they are to emphasize attracting employees closer to their own generation. But when we measured efficiency by workforce age, we saw a marked decline in efficiency in shops with older workforces, while efficiency in shops with younger workforces remained much the same. Finally, we learned that companies with younger and more diverse workforces have lower turnover rates than companies with older workforces and less diversity.

Diverse Workforces More Efficient and Profitable

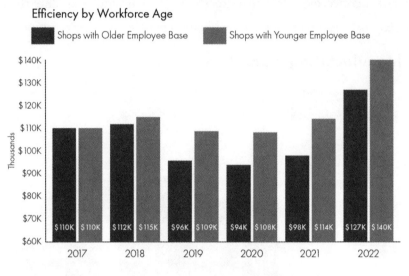

Figure 3.4

Improving Diversity Can Decrease Turnover

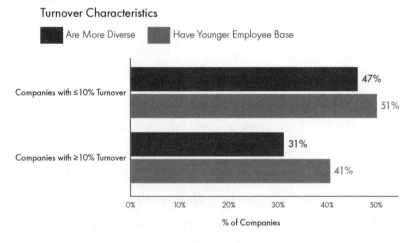

Figure 3.5

So, according to our surveys that take into account the full range of diversity, gender, and a company demographic breakdown that reflects its community and generational mix, manufacturing companies perform better when they're more diverse and have younger leadership.

More Diversity Aligns with Higher EBIT

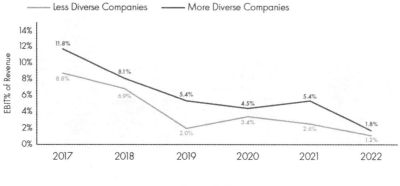

Manufacturing EBIT Trend by Diversity

——— Less Diverse Companies ——— More Diverse Companies

Figure 3.6

When an organization has that, they not only have diversity of leadership, but diversity of thought—a multitude of critical thinkers of all backgrounds with different perspectives on finding solutions and making change. And if you truly want your organization to grow, that diversity of thought is exactly what you need—not more of the status quo.

HUMAN RESOURCES: UNLOVED AND MISUNDERSTOOD

It is an unfortunate fact that many small-to-medium manufacturers tend to view human resource functions as nothing more than an administrative position, usually one that reports to the company's financial officer. Furthermore, that administrator rarely sits on the company's executive committee next to the owner helping to guide labor strategy. Only 50 percent of the manufacturing population across the hundreds of companies we've surveyed have an HR person on staff.

Only Half of Tooling Industry Has HR

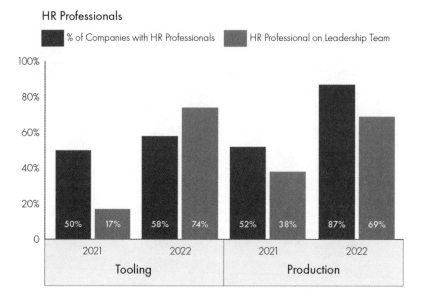

Figure 3.7

The administrator tasked with HR will be responsible for processing training records and tracking job applications. You won't find them sitting down with the head of operations to discuss the recruitment, training, and retention tactics necessary to improve overall company performance. Basically, they'll handle the basic functions of an employment office, but without the strategic mindset. Additionally, if those HR professionals bring up the need to clean the operation and make it safer in order to attract the next generation, most of them are not heard or, worse, told they do not know what they are talking about.

HR Skews Toward Large Facilities

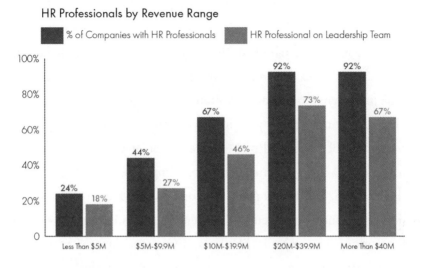

Figure 3.8

Aside from the overall shortage of labor, the manufacturing industry has two significant problems when it comes to finding, employing, and keeping the workforce they need to compete, and in some cases, survive. First, there's no dedicated HR person on staff. Second, if the person with HR responsibility isn't a member of the executive team, then it's a clear indication that the company doesn't value HR as a legitimate function that contributes to profitability.

The latter is somewhat ironic considering that HR is often folded into the chief financial officer's portfolio and viewed as overhead, like sales and finance. There is, however, a direct line between the cost of labor and ROI. The cost of successfully recruiting, hiring, training, and retaining (onboarding) a new employee based on our own work is approximately $18,000. So, if a new worker comes on and then quits in a matter of days or months, you're out of that investment. But if you have a team in HR focused on keeping jobs filled and workers happy, you'll save money. You'll have lower turnover. It's a fact that companies with lower turnover are more profitable.

In many cases, HR is treated as a one-off function and outsourced. Besides the individual who runs the company or the plant, the person who staffs it is the second most important person in the company. It doesn't make sense to outsource a strategic initiative. What is forgotten in that situation is that if you don't work to attract the right people to your pool of talent and make it the center of your labor strategy, that talent will just go somewhere else. Companies such as Amazon, Walmart, UPS, and FedEx pay more for jobs that are, in effect, easier to do. Furthermore, next gen workers tend to get and stay more excited about working for companies like Tesla, Rivian, and Amazon than GM, Ford, or Target because they're perceived as cool and techy.

A MATTER OF MUTUAL ATTRACTION

Unless you're looking to fill a senior management position with a candidate who has demonstrated experience and a stellar track record, you need to focus your recruitment efforts on the current generations of potential employees—Millennials and Gen Zs. You might need a production worker who'll push a button on a machine, but you want a candidate with the potential to learn on the job and develop into a bigger role. These workers have to start somewhere, just like you did. But getting their attention is a different ball game these days, and actually attracting them to your company will take more effort than posting an available position on an industry job site or placing ads in the local newspaper.

First of all, what do Millennials and Gen Zs want? We've already discussed what they want to get out of a job—validation that their efforts have meaning, that their work matters. But they also want a clean, visibly pleasant place to work; an efficient, technology-friendly environment. Who doesn't? They're not going to come to work for you if they can't visualize themselves thriving in your facility. So how do you communicate that potential to a candidate? By linking your labor strategy and recruitment tactics and practices with marketing. You need to *brand* your company in a way that attracts the right workforce as well as selling and promoting your products and capabilities.

It's not the same job market, so why rely on old tools? Today's job seekers are more likely to use Google, Glassdoor, Indeed, Monster, and LinkedIn, in addition to ever-present social media to evaluate company ratings and check employee reviews. They want to know what other people—their peers, current and former employees— have said about the company. They're actively seeking feedback and when they go into an interview, they're looking for certain elements in the facilities and in benefits packages that will "engage" them. They know what they're looking for. Talk to your younger employees and ask them: How can we get your friends to work here? What can we do to make this a great place to work?

It's imperative that you use marketing to pull in candidates just as you'd use it to pull in sales. You need to have an accessible presence on the job search/review platforms mentioned above, in addition to an informative website that provides job seekers with an accurate and attractive picture of your company—what it does, what it stands for, and what it's like to work there. You have to brand your business digitally so candidates can find you. Then you use analytics to determine the effectiveness of your messaging by measuring hit rate and click-throughs.

You have to make sure that job candidates can find you first before they can decide if they want to work for you.

ONBOARDING FOR LONGEVITY

"Welcome to the team. See ya 'round." For many manufacturers, this is the onboarding process. A new hire would show up for their first day of work, run through some basic safety training, be handed a pair of safety glasses, get turned over to a supervisor, and be expected to learn on the job. No one offers to answer their questions or takes the time to encourage them or show them the ropes. Why is anyone surprised when they decide not to stick around?

In poor-performing companies there's very little follow-up with new workers and low levels of engagement and empowerment. Conversely, the best performing companies have proven onboarding programs focused on multiple factors:

1. Comprehensive, structured training prior to exposure to the shop floor

2. Full facility/physical plant familiarization

3. Check-ins (handholding) and regular follow-up

4. Senior team (owner, president) visiting them on a cadence to ask how they are doing

5. Feedback sessions—lunch and learn with ownership

6. Reward system

7. Benefits improvements—vacation, healthcare, pet insurance, etc.

8. Ownership and senior leaders working on the floor alongside hourly labor

Check-ins are particularly important to the assimilation of young talent into the workforce. It's no different from the importance of checking in with your customers; many leaders call associates internal customers and tout the need to treat them as such, so why not start at the onboarding process? When company management, from shift supervisors to general managers and above, regularly take the time to talk to new hires to assess how they're doing from both productivity and employee satisfaction perspectives, they're demonstrating that the new hires are valued. Asking questions such as "You've been here a week, what do you think about your job so far? How do you like it? What can we change or what can we do to make it better? Is there anything you don't like about your job?" gives new employees a platform to contribute.

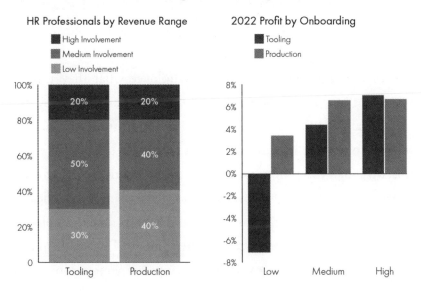

Figures 3.9

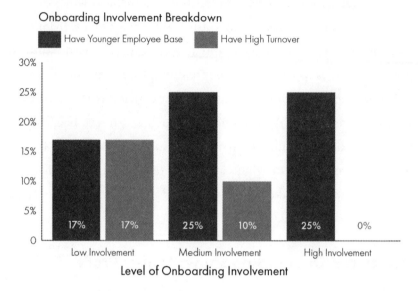

Onboarding Involvement Yields Less Labor Issues

Onboarding Involvement Breakdown

■ Have Younger Employee Base ■ Have High Turnover

Figures 3.10

In the wider view, it tells the employee what they think is important to the operational health of the company. We know that Millennials and Gen Zs are looking for opportunities to make an impact, to provide insight. They want to be able to offer "If we did this, my job or this process would be more productive, more efficient, easier." They want to have a say in a work future in which efficiency and waste minimalization are front and center, and technology is expected to be a tool for efficiency and economy. Engage them from the get-go. Give them a voice. And then watch your turnover rates drop and your profits rise.

THE ROAD TO RETENTION

Making sure that you have a dedicated HR staff, branding your company to attract right people, onboarding them in a way that emphasizes their value to the company—these are foundational issues. If you don't have staff, programs, and processes in place to secure your workforce and keep them happy and productive, you, as a company, are not sustainable. If you have high turnover, you don't have consistency in performance. Having a ready, qualified, eager, and dependable workforce is a strategic advantage because in today's market environment, there are a lot of manufacturers who don't.

Once you've brought a worker onboard, the questions are:

1. What are you doing to keep them?

2. How do you nurture them?

3. Do you have professional development plans in place?

4. How do you treat them?

5. Do you ask for their opinions, their insights?

We've said this before, but it bears repeating: When workers are invested or feel ownership in their productivity and the success of the company, turnover rates are low. The higher your employee satisfaction, the better your labor retention.

Finding the right people was the industry's biggest challenge, and today's post-Covid business climate is still characterized by severe labor shortages. Labor retention remains a serious concern, and just throwing money at it is not the solution. The majority of today's workforce is not motivated by pay scale alone, nor by the benefits package they're offered. Yes, those are important parts of the retention equation, but they're balanced by satisfaction, integrity, and respect. And that's as it should be. All workers need to feel good about what they do, that what they do is valued, and that they're respected and appreciated for their effort. If any part of that equation is out of kilter, they may stick around for a while even if you're paying them top dollar—and with the rising inflation and club dues you're paying to attract people, the minimal level of pay is still going to be fairly high—but, in the end, they won't stay, and that's why these techniques are even more important.

FACING THE UNIMAGINABLE

A group of about a dozen engineers at a modest manufacturing company played the lottery every time the jackpot rose above a certain multi-million-dollar figure. One day they won. And the next day, without warning, all but one of them quit their jobs and walked out.

—

What would you do if a dozen of your workers walked out with no advance warning? How many machines would be idled? How many orders would go unfulfilled? How many shipments would go undelivered? What would happen if key managers in marketing, operations, and finance, along with their top people, were all incapacitated at the same time? How long would it be before the business ground to a halt? How long would it be before you'd be forced to send everyone home?

Contingency labor planning or disaster planning is an area that no one wants to deal with head on. It's unpleasant to think about and painful to process. It's a grim reality that no one wants to accept. But the global Covid-19 pandemic has shown us just how important it is.

Earlier in this chapter, we told you how Omega Tool was able to adapt their day-to-day operations quickly to keep their workers safe while continuing production at the outset of the pandemic. They were ahead of the curve because they were actively working on long-term business strategies, one of which was to figure out how to accommodate the desires of their younger workers (the future of the company), while benefitting from their technological skills. Some of Omega's competitors didn't survive the Covid shutdown precisely because they had no such contingency plan in place to fall back on. Where Omega was able to say to their workforce, "OK. We know how to do this. Go home." Most of their competitors said "Wait, what?"

In a twisted sense, the pandemic has done us a favor. It fast-forwarded the manufacturing sector into a multi-year timeframe during which we were compelled to find different ways to work. We were pushed *way* outside our comfort zones, and it placed an inordinate amount of emphasis on finding creative alternatives to business as usual. The paradigm didn't shift; it broke. The older workforce couldn't envision working without being on-site. The younger workforce, on the other hand, helped us figure out how to be more hybrid, how to run manufacturing as a partially remote operation, and how to up our operational flexibility.

Covid forced us to rethink not only how we manufacture, but also how we do things within this labor environment. All the elements of labor strategy came into play—the attraction of talent, onboarding, training and development, retention, the assimilation of new skillsets and social mores, and the melding of tribal knowledge and legacy expertise with technologically advanced capability and confidence. Now that Covid is manageable, what are we doing to get people excited about coming back to work, or has the manufacturing work environment been permanently altered? Obviously we don't have an answer to that question. Yet.

In the interim, however, we need to remember that bringing together the old and the new is critical to the survival of North American manufacturing. We've observed that if you leave the up-and-coming generation to their own devices, they'll work only as hard as they want to. But rather than judge them on how many hours they work, encourage them and coach them, and then judge them on the product they put out. We've seen that if you pair them with someone who offers them additional pertinent experience and insight, they become more versatile, more flexible, think more critically and ultimately, and more efficient—which is precisely how you drive improvement.

Bottom line: the generational shift is happening. We need to recruit this generation to manufacturing or they will flock to other geographic regions. We need to connect the retiring generation to the new generation and find people who want to impart their expertise and leave a legacy with those who want to learn and do things more efficiently. Companies have to reinvent themselves and be purposeful about bringing in the next generation. Manufacturing's survival depends on it.

HAVE YOU CONSIDERED?

1. What are the demographic percentages in your workforce?

2. How do you respond to labor disruptions? Can you pull workers from one function to address a shortage somewhere else?

3. How do you attract the talent and skilled labor you need? How do you retain them?

4. What are you doing to empower and engage your current employees and new hires in the processes of your company?

5. Are your workers just bodies running machines, making parts and packaging them? Or are they treated as part of the team, solving problems and engaging in activities that make the business better?

6. Has there been any effort to capture tribal knowledge and legacy expertise in your company?

7. How do you attract new hires? Do you have a multi-channel recruitment strategy?

8. Do you have an overall labor strategy?

9. Do you encourage the sharing of legacy expertise with younger workers?

10. Do you encourage your older workers to learn and embrace new technology?

11. Have you asked your youngest workers how they'd improve/change their work experience and why?

12. When you look across your shop floor and through your offices, who/what do you see?

13. Is HR represented on your executive team?

14. Do you have a digital brand?

15. How comprehensive is your onboarding process for new hires?

16. What is your workforce turnover rate? What are you doing to improve talent retention?

17. How did you handle the Covid crisis?

Gather the Data- financial, sales, operations

Commercial
Strategy &
Process

Technology
Road Map

Leadership
Sustainability

Vision/
Strategy

Talent
Strategy

Operational
Experience

Analysis

Assess

IMPROVEMENT PROCESS WHEEL

CHAPTER 4

SETTING COURSE
TO SECURE
THE FUTURE

"It was a stressful time," remarked Jim Kepler, President of Intertech Plastics, referring to a period where the company was losing customers and needing to reinvent themselves if they were going to make it. His comment may easily have been the understatement of that year.

"Intertech was at a transitional point facing a series of developments that could determine the fate of the company. First, we'd been looking to diversify our industrial plant operations beyond the consumer products molding and packaging business that had sustained Intertech over the years but had also proven to be a boom or bust business. After several cycles in which major customers exited for reasons beyond our control and business dropped more than 50% practically overnight, we were seeking ways to insulate Intertech from such dramatic swings. When we automated production to satisfy one of our consumer products customers, we felt that we had no choice in the matter. Then when another customer pushed their production offshore, we were forced to undergo a round of reductions in workforce. We realized that we were working too hard for too little money and trying to build up a successful business in a volatile, cut-throat sector. We had to do something different.

"Medical manufacturing was an industrial sector that our founder believed would allow us to expand our business. So, when we became aware of a suitable, family-owned medical injection molder located not far from our industrial facility, we jumped at the opportunity to bring it into the Intertech family. However, integrating medical manufacturing with our industrial consumer products business proved problematic.

"The fundamental operational and cultural differences, the 'divide' between the two businesses, were significant with resentment building on both sides of the aisle. Employees in our industrial facility felt overlooked, devalued, and robbed of resources. At the same time, employees in the medical facility had strong ties to the company's prior ownership and questioned what Intertech knew about medical manufacturing. The medical operation had an air-conditioned, clean-room production floor with top-of-the-line equipment. Our industrial plant was hot, loud, and antiquated in comparison. Adding to our difficulties, we discovered that our medical acquisition was experiencing quality issues with several major customers.

"Our founder and owner was also looking to step down from his 35-year role as President and entrusted our leadership team to guide the company without him and to safeguard the company's investment in medical manufacturing. He directed us to take whoever we needed from our industrial operations over to medical in order to be successful. At the time, however, we had scaled back our industrial operations to accommodate the loss of a major customer and morale was at an all-time low. Meanwhile, we were hearing from team members at the medical site that we were ruining their company and that we were 'coming in and doing stuff that didn't make sense.' In actuality, the medical business was on a failing course due to quality issues and on probationary status with some of its largest customers. They were running shifts with seventeen to nineteen people doing manual inspections on very small parts. You can have great operators, but if you're inspecting manually and visually, you can't possibly be 100% accurate. So, we knew we needed to come up with a solution in an environment that wasn't automated.

"Around this time, I attended one of HRI's breakout sessions on Sales Strategy at a MAPP conference. We knew we needed some outside help to facilitate strategic planning because of the overwhelming issues we were facing including the departure of our owner. In the past, we'd been strictly tactical and now we needed to think bigger picture and address serious course corrections. We retained HRI to help us develop a strategy to grow Intertech, align all parts of the company to common goals, and set the vision for five years out.

"HRI challenged us. They asked 'What does success look like in five years? Why are you in business? What is your purpose? What's your why?' This was a new exercise for us and we struggled with it. In the end, we went back to our founder's original vision, why he started the business and why we wanted to work there. Finally, we articulated what we believed to be our common goal...'A connection of development, learning, education, employee development and commitment to a greater purpose outside our own success—the success of the next generation of manufacturing.' We decided that we'd be what our founder had always envisioned—a learning organization. We also knew that we needed to achieve a significant stretch of profitability through increased sales and revenue growth as a unified company connected through vision, values, and culture, not two disparate at-odds divisions.

"We set our three goals as:

1. Become a 'best place to work' by achieving national recognition of employee satisfaction based on training and development.

2. Grow the company, be profitable, and align as one company based on culture and value.

3. Brand Intertech and build a reputation based on competencies of excellence.

"Since we'd already automated the industrial side of the business, we believed automating inspection would benefit the medical manufacturing side. But when we started talking about it we got a lot of pushback from the medical team who didn't think we understood their business. So, we pulled together a team to develop an automation cell that would capture the risks we'd identified as critical mistakes in the past. The resulting system won an automation award and received national recognition. It also proved to be a turning point in cultural acceptance as we demonstrated that our team from the industrial division was there to support the medical side of the business, improve operations where possible, and keep them directly involved in the overall process. We made sure they had ownership of the solution.

"Our customers were impressed. Not only did our proprietary inspection automation cell launch flawlessly, but it was also able to detect every single defect mode. When we were challenged to validate it through hundreds of thousands of parts with 100% inspection, it worked perfectly. That's when we knew we had developed a unique precision methodology.

"During our strategic planning sessions, HRI asked us 'What is your differentiator? Can you create an error and watch the system detect and reject it automatically? Can you prove you can do it every single time? People say all the time that they can do it, but they can't prove it.'

"Well, we did. We proved we could manufacture parts perfectly, and that became the heart of our new, one-company brand—Part Perfect—based firmly on our Why."

DRIVEN BY WHY

We're certainly not the first, or even the second or third or fourth, to talk at length about the concept of an organization's Why, or purpose. Popularized by Simon Sinek in his book *Start with Why* (2009) and his lauded 2010 TED talk, the premise centers on the notion that at the core of every successful enterprise, be it manufacturing or some other industry, is a well-defined sense of purpose. In lay terms, it's what gets us out of bed each day, what drives and focuses effort, and what leads to the setting and articulating a vision for the future.

As in the previous Intertech example, we are often retained to help a company with strategic planning. A common problem that we encounter more often that you might imagine is the tendency to set strategy that is too short-sighted. A client will say "We need a strategy" and then will put in place a plan to achieve five things in a year's time frame. The issue here is that these actions are more tactical than strategic. They're not aligned around the vision of what leadership wants the company to be if, in fact, they know what that is. One of the keys to successful leadership is to realize that vision is not a goal nor objective, but that it guides goals and objectives. Leaders and team members must understand the difference between strategies and tactics, and that both are equally important. The fact is that many companies don't really understand the difference between strategic and tactical initiatives. Strategic planning, however, always starts with a company's Why.

Self-examination is rarely comfortable. On a personal level, answering such probing questions as "Why are we here? Where are we going? What's our purpose? What are we trying to do?" requires introspection and close examination of where we've come from. On an organizational level, it requires collective understanding, endorsement, and acceptance. To get a company moving in the right direction, the entire workforce needs to believe in the Why and the Vision first. Then everything that follows—mission, strategies, tactics—can be tied to it or aligned. Many companies think that they're engaging in strategic planning, but are actually working tactical models (what are we going to do this year?) that are not aligned with their Why and their vision.

> *"People don't buy what you do.*
> *They buy why you do it."*
>
> *—SIMON SINEK*

Once when we pushed a client to sit down and define their vision because all they were doing was writing tactical objectives, we discovered that they had no clear idea of where they were headed because they were living in the past. In those situations, how do you become more focused on strategic alignment or positioning? Those companies need facilitation of the process, someone to lead the discussion and encourage them to answer the hard questions, then to rally around a clear vision. You have to break them down before they can build themselves back up and get back to their Why. This can be difficult to achieve without third-party assistance. Sometimes you need someone outside the company to show you what your environment, culture, and/or current state really looks like.

It's never easy. You have to get past the natural defensive reactions to a level of transparency and honesty that peels back ingrown managerial habits and old-school beliefs. When was it ever pleasant to hear...

1. You're not aligned.

2. You don't have a clear strategy.

3. You don't have a clear vision.

4. The benchmark data shows you don't stack up (with your competition).

And then be challenged to figure out...

1. What are you good at?

2. What are you not good at?

3. What are you trying to accomplish?

This is why, more often than not, manufacturers need independent facilitators for vision alignment and strategic development. If you're facilitating your own strategic positioning and planning, chances are good that you're not challenging yourself to approach it with new ideas or from a different tangent. You're not looking outward.

We encounter it often—a typical manufacturer, a second- or third-generation family-owned business, with leadership that is convinced they know everything because they've never been told otherwise. They're overconfident and cocky, usually covering up or compensating for a weakness, and the last thing they want to do is admit to being vulnerable. But that's exactly what they should do. They should be willing to step back and not be the smartest guy in the room for a change. They need to ask their whole team for help in strategic planning and tactical problem-solving, rather than dictating direction that may or may not be embraced by the team.

Instead, we see manufacturers jump into strategic planning and move immediately into determining tactics—the how and the when—without taking the time to step back, sit down, and answer those critical, vision-defining questions first. They aren't thinking big picture or long range—where do we want to be in five, ten, twenty years? What does success look like? What will it feel like? When leadership creates a set of objectives and cascades them down through the organization to drive accountability, but the vision alignment part of the process has been skipped or omitted entirely, then what you're left with are empty mandates that won't instill confidence or loyalty or commitment.

Going back to the Intertech example at the beginning of this chapter, Jim Kepler and his team understood that they needed to get everyone in both their industrial and medical divisions moving in the same direction toward a common commitment to an ideal, before they could focus on improving the company's overall profitability which, as Jim explained, is a result of strategic planning and tactical execution, not an objective. Jim and the leadership team at Intertech realized that they needed a facilitator to help them build a cohesive strategy. As Jim tells it:

"HRI challenged us to define our purpose and our niche first—with brutal honesty. They asked what was our Why and why did customers do business with us? They made us realize that we really didn't have a competitive differentiator, nor could we articulate why customers came to us specifically. We didn't have a strong brand. Furthermore, we were divided. We had a medical plant and an industrial plant... two facilities, one leadership team, but zero alignment in culture and performance and, not to mention, a serious lack of trust. HRI showed us that trust is the foundation of a high performing team, but we weren't being honest with each other. We were second-guessing and under-cutting ourselves about every important action or initiative we were considering. You can't win or achieve excellence with that kind of broken culture. And finally, we learned that open communication is key to holding teams together, as well as moving an entire company in the right direction. As engineers, communication really wasn't our strong suit."

In Intertech's case, before we could facilitate strategic planning, we realized that they needed to boil their collective purpose down to a common denominator. Not an easy task when the company's two divisions were so different in purposes, products, and cultures. As Jim explained, their industrial and medical divisions were an unnatural fit—sort of like a round peg and a square hole. But in stripping away the layers (peeling the onion) to get to what made the company tick, their core values, Jim and his team landed on what would eventually form the heart of their vision, guide their strategy, and lead to the positioning of their company, internally and externally—their "Part Perfect" brand.

"We wanted to make and deliver perfect parts every time. That meant we needed to do business with customers that fit our sweet spot, that required our expertise. So, we targeted the sectors of industry, such as medical parts and devices, that value precision by driving efficiency and process improvement."

HOW'S YOUR VISION AND WHAT ABOUT YOUR VALUES?

True north … Aspiration … Big hairy audacious goal … This is what we're about … This is what we want for our company. However you decide to articulate it, it needs to be your rallying cry. It's what unites your efforts and points everyone in the same direction. And it's leadership's job to drive that vision, reinforce it and make sure that everything the company does aligns with it. Once the whole team is committed to it, then you can build the long-term strategy and tactical executions to achieve it.

Leadership should be willing to say, "Here's what we're going to do—with your help." The point is to develop achievable objectives *together* that will stretch the organization and help it grow. But if leadership believes that they have all the answers and dictates actions rather than solicits company-wide contribution, then the effort is not collaborative. Or if they've been taught to sit back and weigh in at the end of the process, then they'll be perceived as passing judgment rather than participating in the outcome. The best leaders transition seamlessly between consultation, collaboration, and consensus-building. Policy decisions that govern how an enterprise acts, does business, interacts with its surrounding community, etc., should never be a "thou shall" or "thou shall not" kind of edict. They're ongoing, collaborative discussions. The important thing to remember is that when it comes to an organization's vision, everyone has skin in the game.

Vision is what you set your sights on. It describes what you want to be—for generations to come.

For Intertech and Jim Kepler, before they could start building a long-range strategy, they had to re-visit their core values, which were established by their founder and hadn't changed since the earliest stage of the company when they were engaged in the manufacturing of consumer products. But now, they needed to determine if those values were still relevant to their workforce, divided between their industrial and medical divisions. Jim says:

"We sent out a survey to our entire workforce probing the values important to personal life as well as work life. There were 120 data points in all, but the key values turned out to be the same. Then we looked at the company values established in the beginning and asked if they were still meaningful. Did they address what's important to you? Did they apply to who and what we want to be as a company? We retained all of our original values except for profitability, which is a result or outcome, not a value. We added 'diversity' and 'attitude' to include being passionate about what we do and being self-motivated to be the absolute best we can be.

And we rewrote them to emphasize key words supported by a line or two to define or clarify, used hot-button language and memorable phrases to aid retention, and developed icons to visually represent each value."

The refreshed values strengthened Intertech's culture which, in turn, had a cascading effect throughout the company from their employment processes through operations to their customer base. Now, instead of hiring based on resumes and skills, they interviewed first for culture and values and then for skills. They looked for the right "fit" when assessing potential hires. They established a "no assholes policy," meaning that if someone was super-skilled but difficult to work with or had an over-blown ego, they were probably not a good fit value-wise or culturally. Jim adds:

"I'd characterize Intertech's work environment as one of trust and collaboration... very team oriented. You can't have an environment of innovation without those attributes, and it can't be a top-down dictatorship either. We found that when we started hiring for value and culture alignment, it gelled our leadership team and Intertech as one company, not as a company divided by capabilities, capacity or markets."

For Intertech, it's a two-fold formula that governs how they do business with integrity and how they treat people with respect. And it answers the questions:

1. Why do people want to work for Intertech?

2. Why do customers want to do business with them?

Intertech recognized that they needed help developing a strategy that could grow with the company and that a critical part of that initiative was alignment of the team—meaning the entire workforce, not just leadership—to common goals. They also had to adjust their "window" from the short term to the long term.

Instead of setting one-year objectives, they needed to be looking at five years out and beyond.

When we started working with them, we realized that there were at least two agendas (with separate cultures and values) at work within the company—the industrial consumer products division and the medical division. A third, silent agenda was the founder's original vision, why he started the business and why people like Jim were motivated to join the Intertech team. In effect, we called them on it. We asked them tough questions and forced them to consider:

1. What does success look like for us in five years?

2. What's our purpose?

3. Why are we in business?

Although they struggled to articulate it at first, Jim and his team finally landed on the founder's original values and knew that they needed to make those values the center of their goal-setting. They decided that Intertech would always be a "learning organization" or, in other words, a company dedicated to continuous improvement in all activities and ventures, open to new ideas and manufacturing solutions.

STRATEGIC PLANNING AND THE ROAD TO THE FUTURE

A manufacturing company's policies should be designed around their Why. Linking policy with purpose provides a guide to achieving its long-term goals. In 1980, Michael Porter, the Bishop William Lawrence University Professor of the Harvard Business School, defined strategy as "the broad formula for how a business is going to compete, what its goals should be, and what policies will be needed to carry out those goals." The keywords are policies and goals. A manufacturing company's policies should guide how it works to achieve its long-term goals. The specific short-term actions required to achieve those long-term goals are tactics.

When building your strategic plan, it is important to acknowledge the "stair steps of growth." Every business has natural steps of growth that create challenges around revenue and complexity. Leaders need to prepare their business and establish a plan for their business to grow.

Stair Steps of Growth

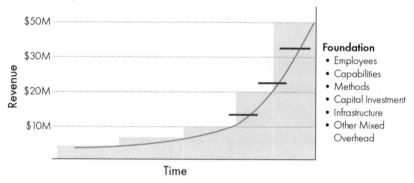

Figure 4.1

When a company's revenue increases from $15 million to $25 million, you need to build the appropriate foundation—do you have the right people, equipment systems, and process in place for example—or you may grow your business but be less profitable and ultimately go backward on the step of growth. Laying the foundation for growth long before you go through it is a critical part of your strategic planning process and should be kept in the forefront of your thinking as long-term goals are established.

There's a variety of strategic planning tools available to manufacturers, so there's no specific cookie-cutter method that can be applied, nor should there be. A company's strategic plan needs to be tailored to the conditions and needs of the company. Furthermore, there are three critical elements or components to securing your company's future: vision, strategy, and tactics. Of the three, vision is the most often ignored. Too many companies jump directly into strategic planning without laying the foundation that will support it or nurturing the culture that will sustain it. Strategy without vision is characterized by short-term results. Strategy without vision is often more of an edict from leadership rather than a connected, collectively endorsed set of objectives. Unfortunately, many companies don't know where to start or how to gain consensus. Leadership may feel that letting go of the reins is not how management is supposed to behave. Sometimes, it's as simple as sitting back and listening to what your employees have to say, or watching how they approach and solve problems without your intervention.

"The first thing we (leadership) had to do was step out of the way," recalled Craig Carrel, speaking for Team One Plastic's leadership. "We had to learn not to be the first one to respond. We weren't unwilling to do it, but it wasn't easy. It took some time before I realized that if I stepped in, maybe my recommendation would be five percent better, but what could we all learn by doing something differently? Eventually, I started thinking along the lines of 'I never thought of that. I'm glad I didn't say anything.'"

Dave Cecchin at Omega Tool agrees: "The old management model where there was little if any team discussion was effective maybe 5 percent of the time. These days, I spend 75 percent of my time listening to what my team has to say... and only 20–25 percent directing them."

But these are hard-won realizations that are not easily embraced by most small-to-medium, family-owned manufacturers characterized by old-school business practices and generational leadership. That's why outside facilitators who understand your industry are essential. They start with the basics—the Why and the What, not the How and the When—back to the roots.

THE SECRET TO IMPLEMENTING STRATEGY

MEASURE WHAT MATTERS

When you've established the high-level objectives, how do you measure success against them? How do you create true key performance indicators (KPIs) that drive toward the strategy and those objectives? There's no set method to determining what needs to be measured but, as a rule, you'll be tracking performance against your objectives to understand how you're progressing against your higher-level goals. The challenge is to make sure you're measuring the right stuff, which will depend entirely on the state of your company.

"We're in trouble!" was the rallying cry for a manufacturer with a weak balance sheet. They formed a strategy to strengthen their financial profile, and a tactical decision to be a cash-only business for a period of time because they felt they had no choice. If they failed, the future of the company would be jeopardized. So, they said that every decision they made going forward would be cash-based and that they'd treat the money as if it was coming out of their own personal checking accounts. In other words, prudently. They realized they needed to change their behavior in order to survive and they could track their success by measuring cash availability and cash flow on an accumulative basis.

But if your objective is to grow your business by 20 percent by making perfect parts in the medical diagnostics arena (like Intertech), how would you measure that? Initially, there would be two things that needed to be addressed right out of the gate: the need to measure, and the discipline around measurement. You have to trust and validate that your KPIs are actually driving the desired behavior. A word of caution, however, don't be fooled if your KPIs are all green or positive, yet you're still losing money. It could mean that your KPIs aren't true KPIs and you're measuring the wrong stuff. The ultimate scorecard is still the profit/loss statement, strength of balance sheet, and cash flow.

There are also times when we measure too much. If a company has twenty metrics, the first thing we'll ask is what do you do with all that data? How do you review it and what are the actions that you take post review? Then what are you going to do when you discover that you're not measuring the right things? There's also a need to delineate between management, leadership, and board-level metrics versus operational and production-level metrics. Your board doesn't need the nitty-gritty on how much raw material you used last week, last month or last quarter, or how many specialty parts you produced for a customer on a specific order. By the same token, not everyone in an organization needs or wants to know how to read and interpret a balance sheet. Share the metrics with the people who have the ability to improve them.

PULLING/PUSHING ACCOUNTABILITY

Once you have the strategy in place and you have the why, the what, the how, the where, and when aligned, then you need a process for team review. This is when you work from the short-term to the long-term—start with the tactical and expand the review process to the strategic level.

In our experience, the most successful companies have daily stand-up meetings to review the most immediate issues of the day. "I've got this die or this mold going into a press, and we also have a customer coming in today" or "We had bad quality last night and we're on containment." A fifteen-minute review of the hot-button issues with the key operations people at the start of the day is crucial to knowing what has to be addressed or solved in the immediate term. A weekly staff meeting involving all core staff is when you can step back and review the week's performance and the next week's job schedule. "How are we doing this week? What containment is in place for that? What do we have on tap for next week?"

Moving from the tactical realm toward the strategic, the executive team would meet on a monthly basis to review the metrics that are fed by the daily and weekly metrics. The quarterly meeting is when progress toward goals and objectives is assessed and reviewed, and the strategic plan should be consulted to make sure you're still on track.

And rather than looking at these meetings as a waste of time, it helps to remember that in the daily/weekly/monthly meetings you're working *in* the business. During the quarterly meetings, you're working *on* the business.

MEETING CADENCE

The failure of many small-to-medium manufacturers is that they tend to over-meet (meet for no reason). Meetings must have a clear purpose and a cadence. A clear meeting cadence will drive both strategic and tactical performance.

CADENCE (FREQUENCY)	PURPOSE
Daily	Immediate (24 hours)
Weekly	Performance/Production Review
Monthly	Metrics Review
Quarterly	Goals & Objectives/ Strategic Plan Progress
Biannually/Annually	Review, Adjust, and Report

Finally, all meetings should be collaborative, working meetings, and all metrics should be important across the entire organization. Understanding the purpose or objective of the meeting you're walking into is critical and all participants should come prepared. Lack of understanding and preparedness wastes the team's time and drives inefficiency.

The following is an example of a metric that is not meaningful enterprise wide. One of General Motors' metrics is to have an exceptional JD Powers score (an external metric), but a GM shift worker responsible for installing a specific part on a car doesn't care about JD Powers, doesn't know what the score denotes, or how it affects him on the job. JD Powers rates the vehicle (essentially a plant-level metric that says 98 percent good vehicles came off the end of the assembly line in a given period). The shift worker who puts on the instrument panel, however, is focused only on his job and not what comes off the line. He may have done a perfect job, but maybe the next guy at his station didn't. Every vehicle on the line goes through hundreds of stations during assembly and errors can be tracked back to the source of the problem. If everyone performs at 100 percent then the finished vehicle is perfect. But what are the chances of that occurring? All the shift worker cares about is if he does his job correctly. If the guy on the next shift doesn't, his station's score will still get dinged on quality.

The highest-level metrics such as throughput, overall quality, on-time delivery, etc., must have supporting metrics that go all the way down the organization to the last worker. When all of these are aligned, the company will perform better as a whole.

One of our clients had a problem with parts falling on the floor. To eliminate the issue, they assigned a monetary value to each part, tracked (measured) how many parts landed on the floor, and translated that to dollars wasted. In effect, they put the problem to their workers in a way they could quickly grasp. Management said X number of dollars were swept up and lost on every shift. By defining the problem in terms everyone understood, they were able to drive behavioral change. They had translated or transitioned a management metric of "why are we leaving this much money on the floor" into something meaningful for the entire organization—missed opportunity.

Another example of a more tactical nature is the Overall Equipment Efficiency (OEE) metric. The problem is if you ask ten people for a definition of OEE, you'll likely get five to seven different answers, so how do you determine the KPIs that will motivate your workforce? A solution is to break OEE into building blocks. Few will know how to move the needle on OEE, but they should know how to impact scrap and cycle time, downtime, predictive/preventive maintenance scheduling, etc., all things that drive OEE.

OVER-ENGINEERING STRATEGY

The strategic process can be over-engineered and often is. Strategy, to be effective, needs to be scaled to the size of the company. When the company grows, so can its strategic scope. A $1 billion company may have a 300-page strategic planning deck because the company has many different divisions and tentacles. But a $100 million company should have a simpler, more straightforward or streamlined strategic plan. It's not about the length of the document. It's about the level of content and the value it provides the team and their ability to utilize it to drive performance.

In any strategic planning process, you have to pick the right questions to ask and the right actions to emphasize. Most people can't really wrap their heads around more than three or four things at once, which means that you need to avoid overburdening your people with paper. Culturally, you have to pick the things that are achievable and set the bar high enough to challenge your workforce, but not so high that they can't succeed. On the other hand, if you just throw a bunch of strategies out there and a year later, nothing's been achieved, then you know that you haven't executed tactically.

Warning Signs & Traps

- Company leadership says they can't do more than three or four things, yet have a 200-page strategic planning deck.

- Too much overhead—Too many people at the corporate level engaged in strategic planning.

- Too much complexity—Too many layers of management involved in planning and not enough in execution.

- Team members aren't allowed to participate—either they're not at the meeting or others talk for them.

- Values are not written or many of them are club dues—without good quality or integrity, you won't be sustainable.

- Plans don't cascade throughout the whole organization—there's a lack of clarity around what each department needs to do to drive the strategy.

- The organization cannot articulate why they exist.

We tell our clients that we'll help them put together objectives and strategies that are simple and will fit on a single page. Their Vision should fit on a single PowerPoint slide and another slide should list the three things they plan to work on. You can devise the most beautiful, elegant strategic plan in the world, but if it's overcomplicated, you won't know how to execute it. And then, it won't make a difference.

TACTICAL EXECUTION AND ATTITUDES

Sometimes the key to executing on strategy comes down to asking the right questions and challenging the status quo. When we hear a client explain "I'm behind on this objective, because we've always done it this way and we don't know how to do it any other way," then our responsibility is to challenge him or her to think and act differently. It's particularly important when it comes to the kind of participation you need in meetings that concern strategy and tactical execution. You should ask "Do we have the right people in the meeting? Are we having too many meetings? Are we talking about the right things? What actions are the teams nailing?" If your teams are cranking away and meeting their timelines, objectives and going in the right direction, then you probably don't need to take up meeting time examining their performance. Conversely, if all your metrics are positive or green, you may need to ask, "Are we pushing ourselves enough?"

Tactical execution depends on a balance of time, leadership, accountability, teaming, and coaching. Everyone involved from the top–down and the bottom–up needs to be clear on the Why, the Vision, the Mission, and Values—as well as the Strengths, Weaknesses, Opportunities, Threats. And then everyone has to understand what you're going to do about these things, how you're going to do it, and last but certainly not least, who's going to get it done. Those elements combined is what makes for a true strategy where companies move collectively toward the future. The most successful have a robust true north *and* the tactics to get there.

HAVE YOU CONSIDERED?

1. What gets you out of bed each morning? What inspires or motivates your people?

2. What do you want your company to be? What do you want it to represent?

3. Has your company's character (culture) changed over the years? How?

4. When asked to evaluate your company's performance, what's your first reaction? How do you feel?

5. If you peel back the layers of your current strategy, what will you find at its core?

6. How would you describe your management style? Are you satisfied with the response you receive?

7. Does everyone in your company understand and share your vision and values? How are they communicated?

8. Do you think the difference between strategies and tactics is understood on your shop floor?

9. Do you ask your workforce for input when setting goals and solving problems?

10. What are you doing with the data you collect? Who reviews it? Do you share it with your workforce?

11. How often do you meet to evaluate metrics and who do you pull into review meetings?

12. Are you positive that you're measuring the things that drive improvement?

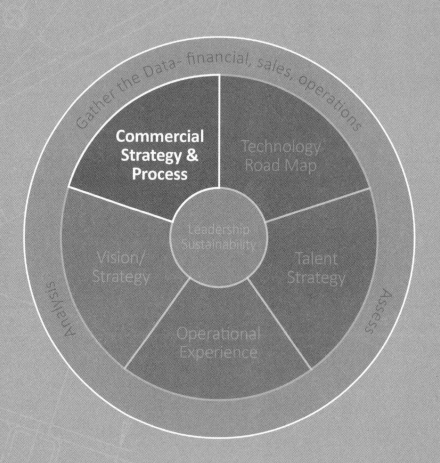

Gather the Data- financial, sales, operations

Commercial Strategy & Process

Technology Road Map

Vision/ Strategy

Leadership Sustainability

Talent Strategy

Operational Experience

Analysis

Assess

IMPROVEMENT PROCESS WHEEL

CHAPTER 5

THE WELL-TUNED COMMERCIAL STRATEGY

Once a client came to us with a problem they described as a bottleneck in their sales pipeline.

"We need to get 700 quotes out the door by a certain date that we know we can't meet. What's wrong with us?" they asked.

Our response was to request a data dump of their quote log covering the past five years, which was confusing to them. They didn't understand why we'd want to look at historical data when their hair was on fire over a very real near-term deadline. We explained that an evaluation of the quote log *would point to issues and vulnerabilities in their sales systems that could be rectified to avoid such issues in the future. After a thorough analysis of the data, we discovered that approximately 80 percent of their business consistently came from a half-dozen customers, yet our client was wasting valuable time quoting a large number of so-called "customers" from whom they'd never won a single piece of business.*

"Why are you quoting these companies when they've never awarded you anything?" we asked.

"Well, you know the day you stop quoting is the day you might get the work," they replied.

"But 80% of your business comes from these six customers and they're prioritized last!" we said.

Rather than help this client get their 700 quotes out the door, we helped them revamp their approach to quoting work. We encouraged them to stop wasting time with customers who were merely price testing because a purchasing department somewhere needed multiple quotes. Instead, we told them to focus on those customers who accounted for 80 percent of their business. Their hit rates on those quotes were high and they won 75 percent of everything they quoted on. Nurture and grow those customers, we said, and let your Business Development team pursue the others.

—

Back in 2010, not only was Team One Plastics quoting every possible job the same way, but Craig Carrel, Director of Sales & Marketing was generating 100 percent of the quotes. He just happened to be the company's President and co-owner, too.

Craig, along with Gary Grigowski, Vice President and co-owner, had learned about HRI through the Manufacturers Association for Plastic Processors (MAPP) and retained our services first for an overall assessment followed by strategic planning. A crucial part of our work focused on the company's commercial strategy or, rather, the lack of one.

"I was quoting everything that came across my desk the same way," Craig explains. "HRI showed me the value of metrics and helped us develop a process for tracking quotes, the number and value of them, as well as the actual hit rates. We weren't doing any of that. We had a couple of heated debates, but they helped me understand that certain strategies I'd employed—pricing strategies in particular—weren't working... we were just bobbing along at the same level, not growing sales at all..."

But now with a more systemic and measurable sales process, Team One has accurate sales data with which to track and analyze their commercial performance and drive key decisions.

"The data showed that we weren't quoting enough based on turnover and hit rate, and that if we wanted to grow by X% then we needed to quote on X more jobs," Craig relates. "HRI pushed us to hire another salesperson so I could focus more on leadership and cultural issues. And then, after we reviewed our numbers over a period of time, we decided to add a second sales guy."

In effect, Team One's prior sales strategy was focused on pro-
tecting and maintaining a level of business rather than actively
growing it. Today, the company's commercial strategy sets
realistic, achievable growth goals that also align with the
company's values and culture.

I t's an unassailable fact that without sales a manufacturer has
nothing. Everything else that we address in the consulting services
we provide is subservient to Sales. Operational excellence, labor
strategies, technology road maps, long-range strategic planning, even
succession planning, none of it matters if you don't have a healthy,
robust commercial strategy. Without sales, you have no customers,
no contracts, no shop floor activity or reason for a plant or, for that
matter, employees. To put it in stark terms, without sales there is no
business.

Fifty years ago in manufacturing it was easier. Sales came to
you. Our industrial sector was strong and our salespeople where less
marketers than energetic schmoozers. We took customers to dinner,
knew their sports allegiances, asked about their kids, and did business
on the golf course. Our customers looked just like us. And there was
enough business to go around, so acquiring sales wasn't the challenge
it is today.

Today, manufacturing's commercial environment bears little
resemblance to the industrial boys' club of the last century. The art of
selling has mutated into something unfamiliar to many of us because
the industrial sector, the sales environment and the people we sell
to are different. Mass quantities of materials come from China and
other low-cost countries (LCCs) and supply is no longer assured.

The buyer profile has shifted from the Boomer-Gen X demographic target to Millennials and Gen Zs, fresh out of school, who may or may not "know" what they're buying and tend to view purchasing as strictly a numbers game rather than an activity based on relationships. Customers now have policies in place that preclude the kind of socializing and gift-giving once considered accepted sales practices. In today's purchasing department, price is often king with supply running second. Getting an inside track on a deal from a longtime customer just doesn't happen like it used to. Customer loyalty isn't entirely a thing of the past. At least, not yet. People still want to do business with people they like and know they can trust, but competitive pressure doesn't always allow for that level of comfort and confidence.

"If your price is higher than someone else's, my bosses will dictate that we go with the low bid despite a good relationship or a legacy of doing business with you." We hear this more and more.

So, what's a manufacturer to do? How do we shift a commercial strategy to address the diverse realities of the current and future market environment?

BACK TO WHY AND LET'S ADD, WHO?

Knowing your company's "Why" is pivotal to determining "Who" your customers should be. If you can't identify your motivation, how can you sell your capability? A manufacturer that's lost sight of its Why will flounder, not just commercially, but operationally as well. So that should be the first component of any commercial strategy.

Manufacturers with poor sales performance share behavioral symptoms that point to the need to do some serious soul-searching. For example, a client of ours with a technically competent workforce found itself off in the weeds so often that they lost sight of their well-established business development process. They weren't thinking critically about operational issues or process improvements. Their leadership team had developed a counter-productive tendency to get bogged down in unsustainable tactics rather than brainstorming lasting strategic solutions. In effect, they were jumping to quick fix-it tactics rather than focusing on solving the root problem. They were bailing instead of plugging the leak.

In this case, the company needed to focus on why they were in business in the first place and what they needed to do before they could dive in to actually doing it. They needed to define and understand their strategic positioning before they could successfully sell their capability. This runs contrary to the old, tactical approach to manufacturing sales (here's what we do, here's how we do it, and here's when we'll deliver it) that's not closely aligned with a company's Why. Ironically, this is not a new idea. In *The Art of War* General Sun Tzu of the Eastern Zhou period wrote "Strategy without tactics is the slowest route to victory. Tactics without strategy is the noise before defeat." Substitute the words "victory" with "success" and "defeat" with "failure" and you have the basic philosophy of an effective commercial program. Words to thrive by and they're only 2,600 years old.

Using ourselves as another example, HRI's Why is to make a difference in North American manufacturing. We wake up every day with that reason for being. The reason why we get out of bed guides how we hire and fire and how we operate. It's our fuel. If a client company isn't aligned with our Why, doesn't "get" our values and our passion, then they're not a good fit for us. This is not empty rhetoric or theoretical navel-gazing. We know value alignment is crucial for success. Manufacturing companies struggle and lose competitive ground when they don't ascertain their strategic purpose and identify the benefit or value it provides customers. And that brings us to the second component of a solid commercial strategy—who's your ideal customer?

DEFINING THE IDEAL CUSTOMER

By definition, an ideal customer pays well, values you as a company, values your product, and willingly partners with you to solve their problems. In other words, they'll invest in the engineering that delivers the value they seek. They'll invest in your Why that delivers the value they seek. An ideal customer won't care as much about price because they value a quality product more and they trust you to produce it for them. With an ideal customer, you don't have to be the low-cost provider. There's still room to negotiate, but if you can wrap your customer's mind around your Why and what that means in beneficial terms, then they're much more likely to want to work with you despite the cost. An ideal customer is aligned with your Why, your vision, and your values.

—

At Team One Plastics, Craig Carrel maintains that the company's Why and their "open book" culture has engendered a level of honesty with their customers that he describes as occasionally brutal.

"Sometimes we get into battles with customers over specifications that don't make a lot of sense and just end up adding cost and not value. We try to catch issues like that at the beginning before we commit. If we say we're gonna do X-Y-Z then we'll do exactly that. It's a commitment and we won't cut corners. We're gonna be transparent because that's how we are, not just with our own team members, but with our customers and

suppliers, too. Let's say we've made an error on a part, maybe used the wrong material at an early stage of production. We don't immediately run to the customer. Instead, we'll spend the time we need to figure out precisely what the problem is, how it happened, how we're gonna fix it and then go to the customer...

"We tell our customers, 'This is how we see things,' and won't sugar-coat it because bad news doesn't' get better over time. They may not like it, but they know that Team One is the expert and they know they need us. In the end, they appreciate our honesty...

"The customer is not always right. In fact, sometimes a customer can be really, really wrong. But if you have a good relationship with them based on shared values, you can convince them to change their approach to a problem and consider your solution. But if they won't change their mind and insist on something that we know won't work or adds cost or is inefficient, we're prepared to walk away from the business. Not every customer is worth having."

UPDATING AN AGE-OLD PROCESS

As we mentioned at the beginning of this chapter, sales is a different game in this new century—less relationship-based and more price, capability, and technology-driven and affected by the same generational divide that impacts labor strategy. Old-school salespeople trying to sell to Millennials and Gen Zs can yield an unprecedented level of cultural and technological frustration. Where a Gen Xer might close a deal over dinner at a customer's favorite steakhouse, the Millennial is more likely to create a comparison spreadsheet and make a buy decision based on the available data, prior experience be damned. How you bridge that divide is a critical element in your commercial strategy.

When you think of a typical manufacturing sales rep, what characteristics come to mind? Personable, affable, competitive, and knowledgeable about his or her company's capability and capacity. Eager to please and driven to win the order. Most salespeople are hard-wired to be outgoing. To some, it's all about who you know and what you can bring to the table. Their game is one of stalking prey and capturing the order. They don't concern themselves with the data of sales. They just want to move their prospect from unqualified lead to quote. These are the "hunters."

Then there are others who have great relationship skills, but lack the thirst for pursuit that characterizes the hunter. They know their clients really well, have taken the time to nurture their business over time. They'll ask clients about potential opportunities and make a concerted effort to maintain a current customer base, but they're not as effective at landing new business. These are the "farmers." (In some cases, these farmers may not be traditional sales positions but project or program engineers who touch customers in different ways and for different reasons.) And both the hunter and the farmer rely on a "closer" to seal the deal. In the small-to-medium manufacturing environment, very often the closer is the owner or a member of the leadership team. This is because the farmer may not be comfortable asking for the order, and the hunter is already running down the next prospect.

Where the hunter focuses on meeting, greeting, and calling prospects, the farmer works at getting to know the potential customer personally in order to maintain a steady flow of business. Neither are particularly data-driven. Because they're wired for relationship-selling, sales metrics, and data analytics based on rigorous feedback are unnatural disciplines for them. But in today's manufacturing sales environment, you can't really build an effective commercial strategy that drives revenue and raises your profitability unless you start with the data needed to assess, evaluate, and shape the sales strategy and develop the tactics to implement it. You can't work an effective sales process unless you feed it first.

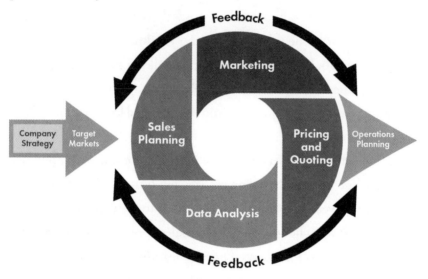

Figure 5.1

When a manufacturer has identified and embraced their Why, their reason for doing business and the values that govern how they do it, they have the foundation for the variety of strategic processes needed to grow and sustain business over the long term. A strategic sales process, such as the example shown, would feature four key areas of focus—sales planning, marketing, pricing/quoting, and data analysis—fed by continuous feedback loops. Company strategy and the identified target market direct the process (input), and efficient operations planning is the result (output). Similar to the Process Improvement diagram we detailed in chapter 1, our strategic sales process shown in Figure 51 resembles a wheel, and its forward progress is fueled by data analysis and assessment that guides and shapes sales planning and marketing efforts, as well as pricing and quoting and, ultimately, operations planning.

DATA IN, PRODUCTION OUT

Market intelligence data will arm you with the information needed to make intelligent, strategic decisions in sales and operations planning, especially regarding the trade-off between price and cost. If Sales says, "We can sell this product for $," but Operations says, "It'll cost us $$ to produce," you have a problem. But if Sales has market intel that indicates that $ is all the market will pay for that particular product, then Operations needs to figure out a way to produce it for $ or less—otherwise you won't get the order. Once again, sales potential and operational capability are linked. If you have good market intelligence, you'll know what pond to fish in and what kind of fish you'll catch. Good market intelligence leads to good sales that fit a manufacturer's sweet spot.

Let's begin with the data because every decision and action in a strategic sales effort relies on it. You need to ask questions such as:

1. What will the market bear?

2. Is this a commodity or a value-added offering?

3. Is this within our sweet spot of capability?

4. Is this a new market need that we can create an adjacency from?

5. How many different parts/products do we make?

6. How much profit do we earn on a per-part basis?

7. What is our volume of parts?

8. Which parts flow through our plant most efficiently?

9. Are there engineering or operational that can be made to be more competitive?

10. Are we using true costs?

11. What parts make the most sense for our company? Which ones align best with our why?

Gathering the data and answering these questions helps to hone in on what your company is good at and shows where and how your sales and operations are intrinsically linked.

BEWARE OF THE BAD SALE

Bad sales, on the other hand, lead to unhealthy frustration and conflict between sales and operations. A bad sale, for example, might be a volume sale. In other words, a large order booked to keep a customer happy, but without consideration as to whether it aligns with the manufacturer's capability and technological competence. When sales says, "Here's a great opportunity," and operations responds with "No, we shouldn't make that because it doesn't play to our strength," that's the kind of honest give-and-take needed for smart sales. The goal is to go from "No" to "Yes, and here's how." The ability of the operations team and the sales teams to delineate between these answers and drive toward the "Yes, and here's how" response is crucial to mitigate the risk of poor sales.

When you weigh the value of your sales consider both profitability and cultural impact. Good sales benefit both. Bad sales benefit neither. Bad sales such as opportunistic sales with no long-term play (no repeat business) and/or high-profile orders (big shiny objects) that don't align with your values or come from a customer not interested in partnering can "cost" an organization. Likewise, sales of products that are manufactured at cost or are difficult for your plant to produce, and orders that are booked solely to keep your plant busy (necessary at times), are not good sales and perpetuate an unhealthy commercial program. When Sales isn't communicating with Operations and over-commits on quality, quantity, price, or delivery, or when Sales is held accountable only to top-line revenue, damage to the company's brand and reputation are often the result.

THE RIGHT SALES REP

Who handles sales in your organization? If your president is also your primary salesperson as is the norm for many small manufacturers, then there's likely a confusing overlap in clear roles and responsibilities. Does a company leader have the time that consistent sales programs require to move new customers through the pipeline from prospect to order?

Consider the following statistics compiled by IRC Sales Solutions.[4] On average, 44 percent of salespeople fail to follow up after just one attempted contact, and only 8 percent follow up more than five times. Ninety-two percent of salespeople give up if they can't make the close by the fourth call. But look at what it takes to actually make a sale: 2 percent of sales are achieved on the first contact, 3 percent on the second, 5 percent on the third, 10 percent on the fourth, and 80 percent of sales are closed on the fifth through the twelfth contact. The fact is sales takes time and persistency. Seriously, what owner can

---

4   "Sales Follow-Up Statistics and Process – The Power of Follow-Ups," IRC Sales Solutions, accessed June 2023, https://ircsalessolutions.com/insights/sales-follow-up-statistics/.

afford to spend that kind of time on sales? Conversely, what manufacturing company can afford to have its senior leadership spend so much time on ginning up business when there's a business to run?

Some manufacturers choose to outsource sales to independent reps. The problem here is that outside sales reps are not usually steeped in your company's culture and the chance is much greater that the business they bring in will be more opportunistic and, as such, not conducive to forming the kind of alignment necessary for long-term customer relationships. But when you hire a dedicated sales person or, better yet, assemble a sales team that also includes an analyst who looks at past and present sales data and reviews market intel, and an administrator who handles the paperwork and a CRM system tied to both email and phone logs so it can track sales-related activity, then your sales efforts are tied to the realities of the market as well as your capacity, capability, and culture. A staffed, experienced, and connected sales department can truly change the future of your business.

BETTER SALES THROUGH MARKETING ASSESSMENT

The sad fact is that for many owners and/or operations executives who are focused on driving throughput, marketing is ambiguous and unquantifiable. It looks too much like overhead and they tend to question "What's our return on it?" It's no great surprise that marketing is often the first budgetary item to be slashed in a market downturn when it should be one of the last. But without marketing identifying and securing business leads is much more difficult. If you don't maintain brand visibility and market presence, how will your customers know you're still in business? In reality, marketing is not often measured properly and marketing data simply isn't put to good use.

First of all, sales and marketing are not the same thing. That's a common misconception, particularly in the industrial sector. Sales is about getting the order to provide customers with the parts they need when they need them. Marketing is about positioning a company in a manner that differentiates it from the competition in regard to capacity, capability and culture. Sales moves product. Marketing builds brand presence which helps identify and attract leads. And without a marketing strategy based on a strong and relatable brand, chances are you won't be able to consistently attract the right kind of sales—the sales that are good fit for your company.

But let's talk about the numbers first since profitability is the ultimate goal. Aside from brand building, a critical element of your commercial or marketing strategy is careful examination of how much you need to do in order to reach your sales goal over a certain period of time. Until you analyze your sales data, you really have no concrete idea of how successful you really are. Let's say you're a $10 million manufacturer that wins, on average over a number of years, 10 percent of your quotes. That's a pretty good average if you want to sustain that performance. But if you're a $10 million company with a 10 percent hit rate and want to grow to $20 million in five years, then math will dictate the potential business—leads—you need to pour into the top of the sales funnel, to generate X number of leads that can be converted to opportunities (quotes) that will result in new sales. In effect, as you move through the process, you're drilling down from your long-range strategic vision through the shorter term tactics that bring in the business.

Figure 5.2

Many manufacturers don't align their marketing goals with their business goals. They don't spend the time needed to identify the number and type of leads needed to achieve a profitable business. Additionally, rarely are marketing strategies aligned with the needs and objectives of operations, human resources, and inventory and supply management, etc. For example, if operations aren't involved at the opportunity stage (RFQ), how will sales know if an order can be executed for a specific price, at the margin needed or desired, and delivered on time?

WEEDING THE GARDEN

Every manufacturer can tell you who the top customers are who represent approximately 80 percent of their sales. Following the Pareto

Principle, this group is likely only 20 percent of their customer base. So, the question is: If 80 percent of their customers account for only 20 percent sales, as a percentage of revenue, how valuable are they to the company? On a chart, we'd refer to the customers who account for only 20 percent of sales as the "tail." Chances are good that the customers in the tail are not strategically aligned with the company. In other words, they're not the best fit.

Find the Story in Your Data

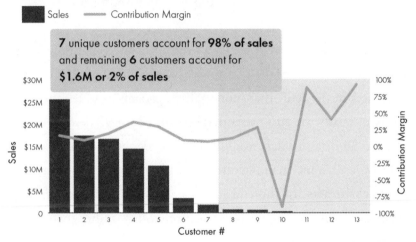

Figure 5.3

Here's an example from our portfolio of business. One of our clients had a total of twenty-two customers, six of which accounted for 92 percent of their revenue. Two of the six were truly huge. This company manufactured thousands of parts. As a percentage of revenue, however, 450-part numbers accounted for less than 0.5 percent and 600 other parts generated less than $1,000/year in revenue, creating significant waste on the shop floor from an operational standpoint. The customers for these parts were in the tail. We suggested that our client review each of these part numbers in regard to how long it took to set up and produce each one and how that, in turn, impacted facility throughput. Initially, it looked like we were assessing operational data, but it was also sales data. They needed to understand that their shop floor was inefficient because their salespeople were selling anything, anytime, anywhere regardless of the impact on throughput, or because over time volumes have gone down, and it has become a legacy business. In effect, it was symptomatic of loyalty to customers at the expense of efficiency. Lots of low-volume jobs requiring multiple setups are more of a drag on efficiency.

So, in order to drive profits, we recommended that they "weed" their garden. Thin out their customer base by raising prices on those customers in the "tail" that monopolized capacity and lessened their ability to satisfy the customers that accounted for 92 percent of their revenue. Some customers wouldn't like the higher prices and threaten to leave, and some manufacturers would say that's no way to treat legacy customers. In the end, it's poor business sense to continue to carry customers that curtail your profitability, rather than forcing them to pay a reasonable market rate.

Over the last few years, manufacturing has become a high mix/low volume business with more and more customers interested in customization. Companies have to be careful not to pursue too many low-volume jobs that have the potential to create chaos in the facility and impact profit. There is definitely a time to take this kind of work (strategic customer with lots of other parts or a new customer you're trying to acquire), but for the most part, you have to be careful not to drive inefficiency in your operations. The caveat is that if you've designed your systems and processes to suit low volume/high mix, then go for it. Unfortunately, many manufacturers try to force both models into the same processes.

NEXT-CENTURY MARKETING

Gone are the good old days when a manufacturer's primary marketing channels were vertical market or trade magazine print advertising and trade show participation. Prospective customers, many of whom are Millennials or Gen Zs, aren't requesting reams of collateral materials such as capability brochures and technical product sheets. Trade show participation is down with fewer and fewer manufacturers footing the bill for extravagant booth exhibits. The digital age has opened new avenues for marketing communications and networking—websites, e-zines, social media, podcasts, blogs. And branding is now a two-pronged initiative aimed at informing and impressing two pools of prospects—new customers (sales) and potential employees (labor).

Today, a manufacturer's marketing efforts and channels need to be purposeful and measured. You need to understand how potential customers are using your website and track the individuals who are following your company on LinkedIn, Facebook, and other social media platforms. Industrial networking, the source of many a sales lead, is moving away from icebreakers and cocktail receptions toward online panel discussions and webinars, and the pandemic served to accelerate this trend with the curtailment of face-to-face selling. We were forced to learn how to use GoTo Meeting, Zoom, Google Meet, Microsoft Teams, and WebEx, among others. When Covid-19 shut down business-as-usual for North American manufacturing, it emphasized the need for our industry to learn and adopt new technologies, new methods, and modes of communication, not just with potential customers but with each other as well. It's a fact that most of the companies that came through the pandemic with minimal losses (or actually won business) during the eighteen to twenty-four-month lull already had strong, active brands with a robust media presence.

A timely example: A ventilator company contacted General Motors requesting manufacturing assistance in ramping up the production of desperately needed ventilator parts. GM's Purchasing Department ran an online search of likely suppliers and found a die-caster in Minnesota that could produce the necessary precision parts and send them to GM's Indiana facility for assembly. The manufacturer was well-branded and search engine optimized (SEO) to appear at the top of the internet search listings.

ASSEMBLING THE RIGHT TOOLS & USING THEM

We've already explained how important gathering and analyzing data is to shaping a strong commercial strategy. You should be gathering hit rate information, number of quotes you generate weekly, monthly, yearly, and how many jobs you land and don't land. You should conduct post-mortems on all wins and losses. What did you win and why, and what did you lose and why. You should look at your target pricing. What will the customer pay and what will the market bear?

> Target Pricing: The process of estimating a competitive price in the marketplace and applying a company's standard profit margin to that price to determine the maximum cost for that product.

The more market intelligence you can amass, the better equipped you'll be to evaluate potential and the price of entry or cost to play in that industrial sector. The questions you need to answer are:

1. How big is this market?

2. What are the market segments?

3. Who are the players?

4. Who are my competitors?

5. Who is already in my space?

6. How do I differentiate?

7. Why do I want to be in this market?

8. What's the price of entry (costs, regulatory requirements, equipment/automation, etc.)?

9. What's my desired outcome?

10. What can I expect to win realistically?

Answering those questions is key to the sales planning process in which you'll build customer-specific account plans and segment marketing plans, run market studies, create pitch documents and other marketing communications to reach your prospects. Finally, you need the right people to perform these analyses and monitor continuous updating. They'll build your sales forecasts and metrics, generate the sales pipeline reporting and the system needed for tracking and managing it.

Armed with market intelligence, you can define your ideal customer, one who is aligned with your why, your strategy, and use the definition as a customer filter tool along with capability, capacity, and logistics considerations to determine whether to pursue a particular prospect or pass on the business.

For example: A salesperson for a manufacturing company in Tennessee uncovers a new business lead that offers entry into a new, potentially lucrative market, but this prospect is located in a rural area far from any major industrial center. What are the questions that need to be answered before a go or no-go decision can be reached? How big is the opportunity? Where are they located? How easy are they to get to? Do they align to our values? What is our current utilization? What's our available capacity? Does the job line up with our skillsets? The answers to these questions and others provide a set of filters to rate the desirability of the customer because the truth of the matter is that not every customer is worth having.

ALL ABOUT THAT BRAND

Your logo or your corporate mark is not your brand. Nor is your slogan. Your brand is why and who you are. It's what your company stands for, why it's important, and why it matters to your customers. Your brand is what convinces customers to do business with you and instills confidence in your ability to do the job right. Often companies led by engineers or financial guys can't get their heads around marketing or branding or why it's important. They don't understand the power of positioning or appreciate that marketing is a crucial tool in the strategic toolbox. And that it's not just overhead, but can be measured for effectiveness.

So where does the small-to-medium manufacturer begin its branding effort? There are a variety of resources and approaches. Unfortunately, many small manufacturers don't feel they have the budget to hire a professional agency. As a consequence, they end up doing a less than effective or thorough job trying to piece together a marketing program with little thought to alignment of these channels to the company's Why, which is where a company's branding needs to start. The result is often a standalone, formulaic website put together by a web designer rather than a dedicated marketing professional, with little or no connection to supporting marketing elements such as trade show participation, social media exposure, online advertising, and public and community relations.

BEGIN WITH WHY

A strong brand communicates differentiation and value to the customer as well as confidence and an attractive culture. Your customers should *want* to do business with you and that should be your brand's objective. If you don't have a clear idea of who and why you are as a company, your brand will be muddy, and any marketing efforts based on it will be unfocused. But how do you build a strong brand if your Why is simply "We Build Parts" as has been the case for many legacy manufacturers. Qualifying that statement with "good parts" or "the best parts" won't cut it because those are relatively empty promises. So how do you identify the idea or concept that needs to form the core of your brand?

In the prior chapter, Jim Kepler, President of Intertech Plastics in Denver, explained why they needed to rebrand during a critical and somewhat turbulent period that was changing who they were and how they operated, both internally and externally. Re-branding the company, as he discovered, was also interwoven with Intertech's need to return to their foundational values in order to build a long-range and perhaps, most importantly, cohesive one-company strategy.

Jim and his team learned that they had to begin with Intertech's Why and build from there.

Investing in sales, if done correctly, will pay you back ten-fold. A client we assessed in Kansas City had been struggling with the president also operating as the head of sales. We advised them to seriously consider hiring a salesperson full time, even though they were only averaging $5 million in revenue. Within one year that same president called to thank us for that recommendation and said it was the best decision he ever made, that the payback had been incredible. Companies are afraid to hire the hunter or to take on the overhead—but, with the right hunter, you can find immense value.

Having a commercial strategy is critical to any manufacturing organization. The old days of trusting your sales team to find leads without accountability are a thing of the past, and the competition is too strong. Data analysis is not just for operations, but sales teams need it too. Combined with operations, sales can bring in work that fits your company, provides strong profit, and most of all aligns with your Why and your values as an organization.

HAVE YOU CONSIDERED?

1. Who handles sales in your organization?

2. What percentage of your sales comes from your top five customers?

3. What do you do and how much time do you spend to keep your biggest accounts happy?

4. Do you know why you're winning certain jobs? Price? Relationship? Quality?

5. Do your customers align with your vision and values?

6. Can you describe your ideal customer?

7. What's the median age of your salespeople? What's the median age of your customers?

8. When you look at your sales department, do you see hunters and farmers?

9. Do you collect and analyze sales data?

10. How much of your business churns or turns over annually?

11. Looking at your customer base, can you identify which are good sales and which are bad?

12. How often do you review and analyze sales data other than revenue?

13. Do you know how much more business you need to book in order to grow your business 5 percent, 10 percent, or 20 percent?

14. Do you effectively use a CRM?

15. When was the last time you reviewed and analyzed your customer list?

16. Do you actively seek market intelligence? Who collects it and what do you do with it?

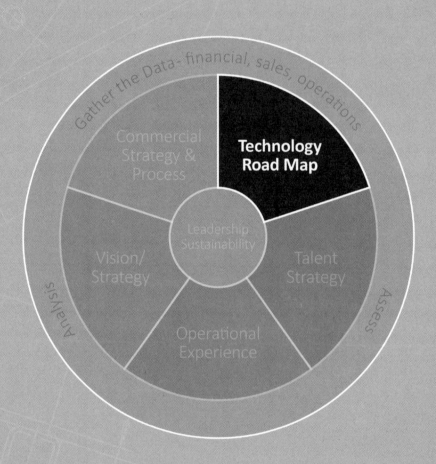

Gather the Data- financial, sales, operations

Commercial
Strategy &
Process

Technology
Road Map

Analysis

Vision/
Strategy

Leadership
Sustainability

Talent
Strategy

Assess

Operational
Experience

IMPROVEMENT PROCESS WHEEL

CHAPTER 6

TRAVELING THE TECHNOLOGY ROAD MAP

There are three key questions manufacturers need to answer when thinking about developing a technology plan. First, what is the current state of your company and what are its needs across the board? Second, how and where are you using technology throughout your enterprise? And third, what do you want the company to look like in the near-term and long-term future?

But even before you answer those questions, you might consider how you define technology and think about how technology impacts our lives today. And then look back at where we were as an industrial society fifty, twenty-five or even ten years ago.

Among its diverse definitions, the most appropriate for our discussion is *technology is the use of scientific knowledge or processes in business, industry, manufacturing, etc., for practical purposes.* Using that definition, technology is pretty much everywhere in our lives, in our businesses and especially in industrial operations.

Fifty years ago, computers were still clunky boxes with cathode ray tubes and typewriter-style keyboards, and portable computers (pre-cursors to the laptop) were the size of small suitcases and weighed between 30 and 50 lbs. and had less than 1 megabyte memory. We used floppy discs and 3 1/2" diskettes to store data (remember those?). To access the internet, we plugged our phones into modems. In office environments we used word processors with limited memory and fax machines to share information and held audio conference calls around tinny microphones in the middle of a conference table. Video calls were expensive and complicated. If a president or general manager needed to speak to his or her workforce, more often than not they shot a video that was then shared in an all-hands meeting. Portable phones were the size of shoes and were often hard-wired into the front seat consoles of our cars.

Fast-forward just twenty-five years to the turn of the century and cellular phones had more computing capability than the computers on the Mercury, Gemini, and Apollo spacecrafts. We routinely conduct business video calls from our cars, airport lounges, home offices, even our kitchens using mobile technology that was unimaginable even ten years ago. In manufacturing, assembly line automation evolved from programmable machining and basic robotics to advanced CAD/CAM, autonomous manufacturing processes and computer-controlled, remotely monitored lights-out facilities. The point is in just the last half century, technology has changed how we live, work, and thrive. And the pace of technological innovation and adoption is not slowing down. It's accelerating. So, the question facing society, all business, and all industry is: How do we keep up? How do we *not* get left behind?

Let's go back to our improvement process wheel.

Figure 6.1

In a previous chapter, we talked about how a thorough assessment of your operations (beginning with the data) will indicate where you need to start—Operations, Labor, Commercial, Vision/Strategy, Technology, or Leadership. A critical part of this overall assessment should be a thorough evaluation of the technology you're currently employing in each area.

Most manufacturers think automation when they think of technology planning, but it's much, much more. There's ERP and CRM systems—end-to-end solutions for operations, labor, and commercial activities—that are continually evolving to offer more control, greater efficiency, better productivity. There are advanced business communications systems that are changing how, when, and where business is conducted, and transportation systems that are edging closer and closer to autonomous shipping and delivery, and most recently AI. The point is that technology and our use of it, especially in manufacturing, is ubiquitous, and the next generation of it, Industry 4.0, is not just around the corner. It's here.

Industry 4.0 and What It Means for Manufacturing[5]

Industry 4.0 or the fourth industrial revolution refers broadly to the cyber-physical transformation of manufacturing. It is a logical extension of Industry 3.0 in which computers were introduced into manufacturing processes. Industry 4.0 connects those computers together and goes beyond the shop floor to the adoption of the Internet of Things (IoT) and the creation of smart factories and digital manufacturing. With Industrial IoT or IIoT, manufacturers are able to connect smart sensors to computers and other connected technologies to work together to achieve levels of automation not previously possible. Not only will humans communicate with machines, but machines will communicate with each other, and data will optimize all processes in real time from design to delivery and eventually throughout a manufactured product's lifecycle.

---

5   ©2017-2022 TechTarget, https://www.techtarget.com/searcherp/definition/Industry-40.

Key Benefits:

- Sensor data enables monitoring manufacturing processes in real time

- Employee training enhancement with augmented reality in machine operation and safety practices

- Avoidance of equipment outages through machine health sensors and predictive maintenance

- Shop floor tied to back-end corporate systems to enable big data analytics to assist in trend detection

Industry 5.0 is already being pushed in Europe, bringing sustainability and environmental considerations into the industrial thought process. It's time to think about how your business fits into the bigger technology picture.

So where do you start? The answer is with the assessment we've already talked about and the data. Once you understand what your data is telling you, then you'll know where you are on the technology road map, so to speak. For example, Level 1 is basic hand tools and Level 5 is fully integrated feedback loops. Then all you need to do is determine where you want to go and how to get there with a reasonable, achievable technology plan.

Show Me The Technology

Automation	Level 1	Level 2	Level 3	Level 4
Software	ERP/MEP	CRM	MS Office	B.I.
Process	Scheduling	OEE	Costing/Quoting	Automated Variance Control
Quality	Manual Instrument	Automated Inspection	Automated SPC	Predictive Control
Human Capital	Attraction	Onboarding	Training	Labor Augmentation

Figure 6.2

A common question faced by the sector might be: What are you doing to accommodate Industry 4.0? Obviously, the answer would depend on the status quo of your organization. A technology plan for a $30 million manufacturer will not be the same as the one devised by a $100 million manufacturer.

Let's say you want to double the size of your business over a five-year span, or you want to become a leader in medical molding. You'd need to plot your technology road map to fit your vision and your long-term strategy, including your labor strategy since you know you're going to have a hard time finding and retaining the right people. It would follow then that part of your technology road map might be to phase in some level of automation so you can reduce your labor need down the road. Maybe you have a quality problem and you need to find/develop/implement process technology to solve the problem before you lose customers or, better yet, use technology to become more predictive and eliminate problems before they occur.

As an example of the latter case, one of our clients had serious maintenance issues. Breakdowns of equipment prevented them from manufacturing and delivering tools in the contracted timeframe.

Adding to their problem was their equipment manufacturers' issues with equipment servicing lead times. Our client asked us to help them build a technology road map around being more predictive. They needed to understand what and when maintenance actions were necessary to prevent breakage to begin with—to predict failure and address it before it occurred. In this case, the technology plan focused on resolving a current operational issue and putting a system in place for the future. They used their data to create predictive models for failure as opposed to running to failure. A good technology road map, in this case, helped them transition from run to failure to preventive maintenance to predictive maintenance.

In other cases, the technology plan or road map could be synonymous with the business plan. For many manufacturers, a Director of Technology and his or her team would be responsible for developing the overall technology plan, which would be more expansive and cover all the departments, activity centers and disciplines in a manufacturing business because technology doesn't stop on the shop floor. It would include IT, ERP, CRM, automation, marketing and communications, labor, quality control, administrative and financial operation, cyber-security, etc.

Think of how manufacturing was forced to adjust and curtail operations during the pandemic, moving many business operations that didn't directly deal with shop floor activity offsite. How did we, as an industrial sector, handle remote work and Zoom/Teams meetings? How did we keep our operations connected? Some companies, such as Omega Tool, were ahead of the curve and able to immediately adjust their operations to protect their workforce and accommodate the need for remote work. In their case, they had already embraced the younger generation's familiarity with advanced communications and computing technology and were ready to implement a new way of conducting business. Another of our clients is less comfortable with the pandemic-mandated change in business-as-usual and has questioned how much they rely on email. Wouldn't it be better for business relationships, they asked, if they simply walked across the hall and had a conversation? That's certainly a valid point. There are times when face-to-face interaction promotes critical thinking, but it can also hint at resistance to change and an unwillingness to fully accept technological innovation despite obvious efficiencies. Ultimately, this is a matter of judgement. There are times when the walk down the hall makes total sense. We have worked with several organizations that just check a box by accepting the technology but that don't push critical thinking, which occurs through direct interaction. This can be done on video calls, but it is more difficult. Best practice is a combination—meeting your organization where they are at in the technology journey.

The shift that an organization makes toward implementation of a technology road map should be directed to answering questions such as the following:

1. How do we better utilize the most important, and limited, asset that we have—qualified talent and their skillsets?

2. How can we enhance the work/shop environment to perform at higher levels, produce more, or do more with what we have?

3. How do we take multiple, but adjacent, job descriptions and transform them to move away from conventionally siloed work responsibilities?

4. How do I better utilize my human capital and soft assets?

5. How can I do all of the above through better technology?

6. What process technology should we consider?

7. What portions of my process can I hand or semi-automate to make incremental improvement

8. Have we mapped every process and challenged its level of efficiency?

9. Is there predictive technology that can help us anticipate problems?

10. Is there predictive technology that assures better quality and zero defects?

11. What type of people should we hire to drive technology thinking?

12. How do we link the younger generation's way of thinking to the experienced generation retiring soon?

The point is that a technology road map or plan shouldn't be restricted to automation of manufacturing processes. It should include everything related to running the business of manufacturing and that's where generational capabilities can be best applied and exploited. It's a given that current and future generations are and will be better at understanding, accepting, adopting, and utilizing technological innovation because they grew up with it. They've never known a world where there weren't laptops and cell phones. They look at technology as a tool to make work life more efficient, eliminate redundancies and waste, and save time. In fact, there are more tools available than ever before. And now, we have the people in our organizations who know how to use them. So, we should tap into their talents, observe what they do and learn from them. Leaders should be able to say "Here's what we're trying to achieve. Can you help us get there with your expertise?" But leaders and old-school managers also need to teach the younger generations to see what they see and how to look at the important functions of manufacturing so they can help develop the right technological tool kit for the job.

For example, today there are powerful business intelligence (BI) tools that link ERP systems with scheduling and other systems, all critical to manufacturing. Ideally, you should be able to pull data from all of those systems into a BI tool to present a snapshot or status quo report to operational management and company leadership that tells you "Here's what all of our data is telling us about the state of our business." The old-school generation (Boomers and Gen Xs) probably wouldn't have any idea how to go about creating such a tool kit, much less use it, but a younger worker taught in a digital environment would.

Where an older worker may resist change because, "This is what I was taught back when... and we've always done it this way," the younger worker's approach would likely run more along the lines of "Why do we do it this way? If we do it another way, we won't need to repeat that action or motion." In effect, they'd solve the problem by eliminating unnecessary repetition. In this way, the technology road map can also serve to bridge the generational gap in manufacturing methodology and the deployment of digital tools. It allows legacy experts to teach the upcoming generations about manufacturing processes while encouraging the younger workforce to bring leading edge technology into the manufacturing environment—while both are focused on improving overall operations.

In the end, all anyone needs to remember (and appreciate) is that technology provides solutions to manufacturing problems and enables enterprise-wide improvement. Technology makes manufacturing companies better at what they do.

LEADING THE PROCESS, DRIVING CHANGE

We don't tell manufacturers what kind of technology they need or where to implement it. That's not our role because every company is different, and every plan should be tailored to meet specific needs and objectives. But we do tell our clients that every company should have a champion driving the technology plan, someone who can read the road map and act as navigator. It's not necessarily the owner of the company or the CEO or COO. Instead, it needs to be someone who can solicit input from the entire company and clearly articulate the cost/benefit argument for investing in technology as a strategic hedge for the future. The technology champion should ask what the needs of our salespeople are before implementing a CRM system; what are the needs of operations and maintenance before selecting an ERP system. And what does HR need? Finance? Quality Control? What do our customers need before implementing new products or processes. Assessing the company's status quo and then its immediate, near-term, and long-term needs across the board is critical, particularly if you're asking, "What do we want to look like in the future?"

ASSESSMENT > NEEDS > VISION > STRATEGY > TECHNOLOGY ROAD MAP

Assign a technology champion or team to drive the initiative and monitor progress. It could be the executive team that sets the vision/strategy that recognizes the need for a technology road map to secure the company's future. But what will that road map look like? It will be different for each company. For example, a company that builds a lights-out facility for the efficient manufacture of advanced plastic parts won't have the same technology needs as that of a medical manufacturer with a zero-defect mandate. So, the assessment of what your company needs is critical to figuring out what will work best for you.

Regardless of who your technology champion is, having the endorsement of top leadership is key. They are the ones, after all, who set and articulate the vision for the company. They are the ones who can take the handcuffs off the technical team and give them license to look for unique technology solutions to fulfill that vision and fund the required investment. Without buy-in from the top, you'll lack the momentum needed to change attitudes and behavior. More importantly, if there's a lack of vision at the top, how will you know where to go with a technology plan? Finally, if leadership isn't sure that customers will value the technological solutions you propose, even if your team comes up with great opportunities that promise to generate significant ROI, chances are relatively slim that you'll get the go-ahead to spend the capital needed.

There has to be a serious commitment to a technology plan. Anything less is a waste of time and money because you won't get an immediate return on your investment. Each multi-year technology road map features a series of prerequisites that, if not met, can hamstring the long-term results. You have to crawl before you walk and walk before you run. And that's where a lot of the challenges that can derail a plan reside... in the prerequisites to full implementation.

WHAT HOLDS US BACK

Let's say that a multi-generational manufacturing company has a legacy vision that loosely defines what they want to achieve in the next two to five years. They have a strategy for the business, but for the past two decades, they've produced similar parts with existing aged equipment. In effect, they've just stuck with the status quo. And because they don't have a technology plan, they'll just keep doing what they've been doing over and over again. They just repeat modestly productive practices when what is needed to propel them into a profitable future is a paradigm shift. They need to realize that while they've been marking time, science and technology have advanced and now can provide them with real opportunity to take their company to the next level and beyond. The old ways were good, but the new ways are exponentially better. The challenge is how to overcome such resistance to change and old, stuck-in-the-past attitudes.

Changing how you think about the manufacturing business isn't easy, nor is determining how narrow or how broad in scope your technology plan should be. But what *is* a fact is that those companies that invest in new technology, specifically in automation, experience that kind of improved efficiency. However, manufacturing technology is not offered in an off-the-shelf digital product line or in plug 'n play systems. You can't just click on "Buy Now" and have a technological solution delivered and working in your environment in a matter of days or weeks. There are foundational steps that you have to take first. Where do you want it to take you? What do you need it to do? Then selecting the right vendor and the appropriate integration and training plans are obviously key to its acceptance and effectiveness. What kind of operational disruption happens when you implement technology your workforce isn't familiar with? And will your customer value the benefit the technology delivers whether it's quality improvements or reduced time to delivery? More to the point, will your customers pay for the added value that technology delivers?

Rather than thinking of technology plans as buying and installing a newer piece of equipment that will magically change the way you manufacture products today, the paradigm shift is in considering how will that new technology better the products you make today, and better the next generation of products? A forward-looking technology road map can present a chicken-or-egg conundrum too. For example, in plastics-manufacturing advanced materials require special handling regarding temperature control and storage and new systems take time to ramp up to full capability. Are you willing to invest in it understanding that it will run initially at a lesser capacity, or will you wait until the customer opportunity is on your doorstep and then try to catch up?

In reality, you can't just buy X piece of equipment and have it magically do what you need it to do. You have to buy the right piece of equipment, set it up, manage your plant layout, train your workforce, and then start pushing the limits of what it can do. The fact of the matter is if you don't have a technical team that understands the full capability of the equipment, how it fits into process flow, the underlying technology, and how it aligns with your business vision, then all you're really doing is running a shiny new piece of equipment and doing the same thing you did before.

Unfortunately, a lot of people in manufacturing don't know what they don't know. Unless you're open to new options, new developments, and innovations you won't know what they can offer you. If you're too set, too comfortable in the routine of doing business the same way over and over, you'll never learn about the potential out there. But once you do, the trick is to keep an open mind. At some point, you need to ignore your resource constraints... at least in theory. Ask yourself what's possible in terms of cost and labor. What might be possible three, four, five years down the road might not be possible now, but what will it take to get there? When leaders' vision cast what they want the business to look like in the future, their team can start identifying specific applications for the appropriate technologies, the resource needs for implementation, the vendor pool to draw on, etc.

But another challenge is overcoming the misconceptions of what technology can actually do. Many people are leery of the latest and greatest technology because they don't truly understand it. As a result, the technology isn't trusted to do the job that it is, in fact, designed and engineered to do. If a system is intended to run by itself and monitor its own operations in real time without human guidance or intervention, a common response is to insist that someone needs to be watching over operations... just in case. There's a preconceived notion that a human will need to make a judgment call, adjust a setting or a process, or be there to flip a switch if necessary. As a result, the same labor inefficiencies are left in the process that the system was designed to eliminate. Workers fear being replaced by machines. The point is, however, not to do more with less but to do more with what we have. Remove the repetitive operation that can be relegated to advanced technology to free up labor to do the jobs that machines can't.

But in the end, the most insidious roadblocks to successfully implementing a technology plan are mired in basic human nature. When the older worker is convinced that the organization's tribal knowledge is what separates them from the competition, that they can do it better because they have more experience or because it's the way it's always been done, then they're blind to the opportunity. When people start considering what's possible, the first thing they list is what's not possible. They revert to "We can't do this, we can't do that... we'll have to do this instead... and that's why it won't work." Instead, they should ask the question "What do we want this to look like?" and then search for the solutions that will get them to that state.

One of our customers addressed many of these technology plan challenges more than a decade ago when they built a lights-out facility designed to run with no one physically in the building. They started with four machines running three shifts, with minimal

159

human oversight on the first shift, and running autonomously on the second and third shifts. Although the equipment components were mostly standard and they didn't need to create anything unique for what they had in mind, the concept was new and how the machines were laid out within the facility was a completely different animal than the routine manufacturing conducted in their legacy facility. It was how they put it together and then made sure that the jobs selected (extended runs to decrease changeover set-up labor) had the highest likelihood of success that was the step-change in their operations. Today, twenty-three machines run sixteen to twenty-four hours in their lights-out facility. And with unique temperature-controlled materials (resins) handling, conveyance and box management systems that rotate automatically based on cycle count, no one needs to be onsite to physically move the finished products.

> "We didn't want to fall back on what we were familiar with just because we thought that was the way it had to be done. For a lights-out facility, we knew we needed the systems to talk to each other and there were a couple of failure modes we had to consider. We wanted to make sure that if a machine failed it would shut down appropriately, access auxiliary equipment, and then communicate its status to the home facility. Now this was before Industry 4.0 and IIoT became industrial buzzwords, so we had to string our own logic together to make sure that if machine A went down then machines B and C would respond so production wouldn't grind to a halt. While it wasn't nearly as technologically advanced as true IoT (Internet of Things), it was a solution to the challenge of 'What if no one's around to flip a switch?' We figured out how to make the switch flip automatically, how to tie the equipment together, and developed the basic programming for the standard controls. It really all boiled down to understanding

what the critical failure modes were and then designing the system and facility with the right countermeasures in place to reduce the likelihood or the severity of those failure modes."

—GENE MUSSEL,[6] *FORMER SENIOR MANAGER, SUPPLY CHAIN, STRATEGY & ANALYTICS PLASTICS COMPONENTS, INC.*

Intertech Medical, which we've talked about in other chapters, is a good example of a company that moved into a manufacturing sector characterized by a set of requirements outside their legacy comfort zone. They needed to change their mindset before they could approach and solve the problem. And then they decided to apply technology and new ways of thinking for the betterment of their company as a whole while increasing their customer's satisfaction. The result was an innovative automated inspection cell that eased their labor burden while ensuring that their medical device manufacturing output was defect-free. In other words, Part Perfect.

ROOTED IN STRATEGY

Your technology road map, just like your labor and commercial strategies, needs to grow out of your foundational vision and strategic planning process, whatever that may be. You know that the answers to questions such as "Where will our growth come from? What will our company look like? What will be our customer/product mix in the future?" are fluid. Unless you've committed your manufacturing business to doing more of the same, there will be changes in how you

---

6   Full disclosure: Gene Mussel now works for HRI as part of our consulting team focused on business strategy and technology integration. He utilizes data analytics to identify opportunities for business and operational improvements, as well as implements strategies to improve efficiency and profitability.

conduct business and how you compete, and it will be critical that you review and refresh your strategy on a regular basis. (We recommend every six months.) Tomorrow's customer may have different requirements and/or expectations. If you're to meet them and grow into new markets, you'll need to evolve with them. And just as you review and adjust your long-term strategic plan, you'll need to do the same with your technology road map to accommodate rapid change and ground-breaking innovation.

The reality in today's manufacturing world is that someone will catch up and pass you by if you don't identify and embrace the technology that will help your business continuously improve itself.

HAVE YOU CONSIDERED?

1. Do you know how you're utilizing technology in each area of your company to improve business performance?

2. Are you keeping your business systems current with the latest software suite releases?

3. How interconnected are your business systems? Can you share data across platforms and departments?

4. Who is responsible for the technology road map in your company?

5. What is your workforce's attitude toward the adoption of new technology? Do they welcome improved capability? Do they fear replacement?

6. How comprehensive is your technology plan?

7. Who serves on your technology planning team?

8. Do you encourage your workers to train up on new technology? Do you provide them with the technical education necessary for proficiency?

9. Do you think of technology as a necessary evil or the wave of the future?

IMPROVEMENT PROCESS WHEEL

CHAPTER 7

SHIFT CHANGE— LEADING INTO THE FUTURE

"Is she ready?"

When Jenn Barlund stepped into her leadership role at Falcon Plastics, she had already demonstrated a healthy work ethic, a deep understanding and appreciation for the business, and a willingness to address issues facing the plastics manufacturing sector. Not only had she learned the family business at her father's and grandfather's knee, but she had worked all over the company in various departments in all three plants. Moreover, she spent a year embedded on our data analytics team working, listening, learning, and immersing herself in a variety of different manufacturing businesses and in the chal-

lenges of the sector. Although manufacturing was in Jenn's DNA, she wasn't sure at first if continuing the manufacturing legacy of her family was the right career path for her. She was (and still is) young. She was female in a traditionally male environment. And she was family, which she felt was a two-edged sword. She knew that her performance as a leader would be judged initially by those stereotypical, but erroneous, misperceptions. Too young to lead. Woman in a man's world. Inherited the leadership role instead of earning it.

We knew she was ready. But we also knew that moving her into leadership was nothing less than a generational step-change for Falcon Plastics. We encouraged them to take that leap of faith because we felt Jenn had already proven she was more than capable, strategically minded, forward-thinking and willing to make the hard calls. She possessed the wisdom and the temperament that characterize leaders of consequence. But her predecessors at Falcon Plastics still weren't completely convinced.

"Is she ready?"

"Were you?" we asked.

—

Kyle Klouda, President of MSI Mold Builders gives several examples of the leadership differences between generations, particularly in communication and work environment.

"I'd say that the biggest challenge facing young leaders in our business is the need to influence older team members who have decades more experience. The question I'd ask is 'Can you manage up?' When I was embedded with HRI at the *beginning of the pandemic, working on virtually assessing manufacturing clients, we spent a lot of time talking about that issue. How do you influence people when you don't have managerial power? How do you lead without authority? The beauty of consulting is that you are able to influence without authority. You've been entrusted to assess a business and, if necessary, tell them that their baby's ugly or maybe not as pretty as they think and help them understand why. Then how to you get them to change? As an outsider, the issues may seem obvious. But to an insider, they may not be apparent or seem as important as others that are more immediate. That's where I think young leadership has an advantage. We're interested in efficiency and optimizing processes to reduce waste of any kind. And in manufacturing, there's a lot of low-hanging fruit that older generation team members don't necessarily see because it's always been that way.*

"The older generations [Boomers and Gen X] weren't inclined to tell team members what a great job they were doing. In their minds, they'd just be thanking them for doing the job they were supposed to be doing. But as a younger person in this industry, I can see that the younger generations are interested in feedback on their work performance. They want two-way communications and to feel that what they do is impacting the organizations that they are working in. They also place more importance on the state of their work environment. In contrast, older workers came in, worked their shift, and didn't consider that shop floor conditions could affect the quality of their work.

"As a younger leader, it's easier for me to talk to my team. They know they can ask for feedback in order to make changes

167

where necessary. I can be more flexible when it comes to bettering our work environment to increase productivity, job satisfaction and talent retention."

Kyle also notes the differences in how the concept of hard work has evolved.

"Where an older manager may focus on how many hours you're working, using it to determine how hard you're working and gauge your commitment to the company, a young leader would likely be more focused on how effective you've been at your work tasks. I'll want to teach everyone how to be that effective so we can get better as a company and benefit collectively."

Jenn Barlund and Kyle Klouda are both third-generation heirs apparent in their companies. Neither were completely sure that continuing their family legacies in manufacturing was the right career path. But now, after a couple of years of exposure to a wide variety of sector issues, as well as analytic and operations experience, both are leading their companies into a manufacturing future that is quite different from the one experienced by their fathers and grandfathers. It's faster, more efficient, more competitive, and far more culturally focused. It's also technologically advanced.

INFLUENCE VERSUS AUTHORITY

How do we define leadership? What characteristics and behavior separate good leaders from ineffective ones? At its simplest, leadership is described as an influential power relationship in which the power of an individual or group promotes action or change in others. Within that definition, there's plenty of room for qualification. However, the phrase "influential power relationship" implies a kind of leadership that doesn't necessarily feel inclusive or collaborative. Executive leadership is often understood as being authoritative—the kind of bosses who rule with an iron fist, whose mandates are final and not open to interpretation. That is not to say that all authoritarian leaders aren't effective or "good" leaders.

Dave Cecchin at Omega Tool, whom we've mentioned earlier, is considered an effective, authoritative leader. He's hungry, proactive, always looking for the next opportunity to improve operations and grow the business. The difference is that rather than telling his people what to do, he influences them with questions, challenges thought processes and, through consistent encouragement, forces his people to think critically and creatively, while using all available sources of data and business intelligence.

Today's manufacturing leaders, young and old alike, are finding that leading their businesses in the twenty-first century is a different game. Advanced technology and the ever-accelerating pace of change, and the cascading impact of unforeseen events such as the Covid pandemic, have created a need for extreme flexibility and agility. Manufacturers need to be able to bend and adapt to business pressures such as labor issues and global supply chain constraints. And they need to be nimble enough to react quickly and decisively to whatever comes down the pike both good and bad.

These abilities are second nature to the generations raised and educated within the digital age—digital natives, so to speak. Relying on technology is second nature to them. Access to information has always been at their fingertips. And each new version of an app or an automated process that will heighten efficiency and reduce waste is eagerly adopted. Older leaders are more cautious when faced with such paradigm shifts and slower to adopt new processes, technology and thinking. They need to thoroughly understand the benefit before they'll embrace change. The question is then, how do you guide a leading-edge workforce while retaining your company's legacy of experience, its tribal knowledge?

The answer, as Jenn and Kyle and other next-generation leaders are finding out, is to lead through influence and collaboration, not do-as-I-say authority. And the keys to that are cultural unity or alignment, open communication, inclusiveness, understanding, and vulnerability.

THE CHARACTER OF LEADERSHIP

As a rule, the leader of a company is the chief steward of its culture. When the company is family-owned, the culture stems from its heritage—its legacy—but is not necessarily tied to generations-old practices. When the company is employee-owned, the culture is more of a commitment to and reflection of the vision and values of the whole. But, as in any business, if you have weak leadership, the company's culture will become unfocused or watered down, as will your "why." Good leaders live their company's culture and actively work to strengthen it.

Taking an example from outside the manufacturing sector, a CEO terminated an employee only four days on the job because of how this person was treating his fellow teammates. During onboarding, the new hire had enthusiastically endorsed the company's culture of equity, inclusion, and respect yet, once on the job, failed to behave accordingly. In effect, he talked the talk but didn't walk the walk. His lip service didn't work, and the CEO said, "We don't need that on our team."

Hiring and firing to your company values is not always the easiest thing to do, but it's incredibly important to the maintenance and protection of workplace culture. You must be willing to analyze your own values and vulnerability. It requires significant introspection both as a company and a leader. Leaders who are admired generally exercise good judgment, but when they stumble, they also have the wisdom and humility to course correct. They have the ability to assess the situation from a practical standpoint and dissect it to determine the necessary course of action. When things go south, they bring to bear character, productive habits and healthy team-oriented values to overcome the challenge. They lead by example.

THE TRAITS THAT MATTER

What made an old-school leader great back in the day is not necessarily the same set of characteristics that you'd expect to find in today's young leaders. It's a different manufacturing environment, and today's leaders were educated quite differently and are conversant with technology once only imagined. There's a whole new set of leaders coming up with new tools, capabilities, and expectations.

Furthermore, the global manufacturing industry is changing at an ever-faster rate. The need for organizational flexibility has never been greater. Leadership that embraces change, that can flex to accommodate market fluctuations, labor challenges, and supply chain pressures, not to mention rapidly evolving technology, is best positioned to meet the future. The only constant today is change. So, good leaders look at the shifting playing field and ask: What do we know and what have we learned? Smart leaders never stop seeking understanding and never stop learning. That's a character trait, among others, shared across the generations.

From Trait to Practice

In the leadership forums we host on a regular basis, we ask participants to list what they believe to be the most important leadership traits. And then we examine and refine them into cultural and organizational behaviors. Participants generally list trust, integrity, decision making, consistency, honesty, transparency, inspiring, and empathy as the traits they hope to see in leadership. This is the "group think" answer. We encourage them to apply these traits to desired actions and behaviors that reinforces the company culture and ensures sustainable success. The resultant list looks more like this:

1. Self-aware and prioritizes personal development

2. Focuses on the development of others

3. Encourages strategic thinking, innovation, and action

4. Ethical and civil minded

5. Practices effective cross-cultural communication

6. Vulnerable

In effect, we're now looking at not just desired leadership traits, but what is necessary to turn them into sustaining leadership practices.

BEWARE THE ACCIDENTAL VALUES

Leading by example can also have counterproductive consequences when a leader's habits or behaviors run counter to a company's stated values. Employees look at leadership for clues on unspoken expectations. For example, one leader may like to spend time at the plant or in the office after hours or on weekends when its quiet. Maybe that's a time for introspection. But be aware that when you come in every Saturday and stay until 1 p.m., you may be inadvertently setting an expectation that other team members (particularly in leadership) to do the same. You may not be aware of it, but it's been noted by someone else. So, the comment "I was in the plant on Saturday and I noticed …" carries the insinuation that something happened that team members probably should have known about. Whether it was intended it or not, it puts your team on the defensive.

VULNERABILITY AS A LEADERSHIP STRENGTH

Vulnerability is often perceived as weakness, but it's not. The ability to realize that you do not have all the answers, and that you may not be the smartest man or woman in the room, does not weaken a leadership position. It makes a leader human. And that lack of arrogance actually builds trust within a team.

This concept of vulnerability also means being real and approachable. Many leaders have been taught what it is to be presidential, often meaning sitting back and listening after weighing in with their opinion first for "maximum effectiveness." This is, in fact, not the most effective, and certainly does not foster collaboration; the best leaders are vulnerable—meaning approachable, real people who are mostly willing to make mistakes and even have a dumb idea during brainstorming every now and then.

Having the self-awareness to admit being wrong but willing to learn and seek understanding makes a leader approachable. When an individual strives to coordinate and bring together the right people to solve problems, they're investing in the company's future success through stakeholder engagement. Conversely, an upwardly mobile individual who grows more arrogant as they progress up the corporate ladder may assume they have all the answers upon reaching the top rung. They may feel they've earned the right to dictate rather than collaborate. More likely, they've lost the support and confidence of their team and the company will suffer in the long term.

Why do we do it that way? Is there a better way to do it? The good leader asks questions, but always supports, empowers, engages, and motivates all team members. Their objective is to get them moving in the same direction toward recognized and common goals with clearly stated objectives that are realistic, fact-based, and achievable.

How did we get here? Where do we need to go? The effective leader knows how to extract information from people and other sources and then demonstrate how to use it for the benefit of the company. Key to that is thorough understanding of the situation at hand, the status quo, and its contributing factors. Once understood, behavior can be influenced, and actions can be tied to strategy and tactics—where the business needs to go and how you propose to get there.

What do we need to do? When the ground rules are set and the expected/accepted actions needed to fulfill the objectives are clearly defined, then the group is aligned with the leader and results should be free of surprises, good or bad. Good surprises generally mean that forecasted results were overly conservative. Bad surprises stem from ignorance of contributing factors, lack of understanding, or consideration of consequences such as overdelivering on a program or project to the extent that profits go unrealized.

GROUND RULES ARE VALUES; EXPECTATIONS ARE OPERATIONAL GOALS

When a leader is realistic about both and communicates them in give-and-take terms that people understand and accept, then everyone involved has a stake in the game. That's collaboration, and it's not always easy. Leadership must be able and willing to have the hard conversations and ask their people what they're thinking, how the action or situation affects them professionally and personally, and how they feel about it. This two-way communication is critical to the success of any initiative, project, or program. A team member must feel that they can speak freely about their concerns, that they won't be judged for their questions, and that they can raise a red flag (see something, say something) when necessary—hence, being vulnerable and not having all the answers.

Ground Rules
This is how we operate.

These are our values.

This is our strategic vision.

Expectations (RACI—)
This is what's needed.

These are our roles and responsibilities.

*Responsible, Accountable, Consulted, and Informed

WHAT'S THE PROBLEM AND WHAT ARE YOU DOING ABOUT IT?

The leader that understands the difference between empowerment and enablement listens and understands without removing an issue from the team's level or responsibility. They don't remove the responsibility from the team member and hand it off to someone else for solving. "Yes, I'll help you. What have you tried? What was the result?" are questions that challenge the team member to reconsider the issue while encouraging the effort. The point is to get them to come at the problem from a new direction, view it with a different perspective, and try something they hadn't thought about before. This encourages creative problem-solving and critical thinking. Sometimes a team member seeks validation for attempting to trouble-shoot an unsolvable problem. But if they're requesting help because they've already tried a series of things and now they're stuck, they're actually asking for assistance, new thinking on an issue. Or they want your slant on it (new eyes and ears). In those cases, good leaders ask them to help understand what they've tried and analyze what actions they've taken. They don't take the problem onto their plate, and they avoid the hero syndrome—meaning rewarding those who don their superhero cape to solve the problem. That's reacting and responding—not leadership.

In a collaborative environment when workers come to leadership with what they've done or tried, they're asking "What do you think? Am I on the right track? Am I missing something?" rather than throwing their hands up, abdicating responsibility and expecting leadership to say, "I'll take care of it." The good leader doesn't handle problems, they address challenges. Good leaders challenge their team members to own and solve their problems both individually and as a group. And that promotes accountability.

We have no accountability in this company, so how can we do anything? To be clear, that's just an excuse for why things aren't getting done. What is meant is that no one or group is taking responsibility for their actions or consequences. Unfortunately, "accountability" is rapidly becoming an overused term in business and, as such, is losing its meaning. So, what should it mean to you as a leader and to your workforce? Simply put, accountability is how people take action and get things done.

So Where Does it Start? LEADERSHIP

Figure 7.1

The accountability ladder is a path or journey that can move an organization from inaction to proactive. It is leadership's job to encourage his or her people to eliminate the bottom rung "victim" behaviors (Blame Others, Personal Excuses, "I Can't" and Wait & Hope) and move to the healthy and productive rungs at the top (Acknowledge Reality, "Embrace It," Find Solutions and Make It Happen). But when faced with an employee or team that seems mired at the wrong end of the ladder, leadership needs to determine why. Do people really not want to be responsible for their actions or is something else the root cause of the lack of accountability? Most people want to do a good job. No one wakes up and says to themselves "I'm going to be a jerk at work today." No one says, "I don't want to succeed." Some people don't want to be held responsible in case something doesn't work properly or it flat-out fails, so they adopt a

Teflon attitude (nothing sticks to them) and offer excuses (It's not my fault, I did everything I was supposed to do) instead of trying to determine where and why it went wrong and finding or building a solution. That's when leadership needs to examine how failure has been addressed historically within the business. Has it been treated as a cause for punishment, humiliation, and blame-throwing, or has it been used as a "teaching moment" in which everyone learns?

Promoting accountability is similar to establishing ground rules and setting clear expectations. The more transparent, honest, and open a leader's communications, the more likely their people will take pride in their work and own it.

LEADERSHIP AND THE GENERATIONAL DIVIDE

At present, there are four generations working in North American manufacturing. The Baby Boomers who remember rotary telephones, manual typewriters, black-and-white television, and basic, numerically controlled automation are nearing retirement. The Generation Xers are more technologically sophisticated, but still more analog than truly digital. By contrast, many Millennials and the Gen Zs are digital natives who have never known a time when there weren't wireless communications, portable computing, and the internet and near instant access to information of any kind.

As you'd imagine, leading a workforce with such disparity in age groups and technological ability is not an easy task. Leaders may need to address the different generations with varied skillsets in their workforce in different, nuanced ways. The acceptance issues, job stresses, and resistance that bubble to the surface when time-worn and trusted processes migrate from manual to automated or digital control cannot be swept under the table. Nor can we ignore that when Bob (who's worked for the company for decades) retires, years of experience will walk out the door with him, perhaps leaving the next-gen workers in his department without a valuable resource. How does leadership preserve that tribal knowledge and communicate it to those coming in behind? How do you link the practical expertise of the older workers with the technological savvy of the younger ones?

Acknowledging the innate differences between generations is key, but finding commonality and encouraging understanding is more important. The older generations were taught to be methodical, process oriented. They knew the "tricks of the trade," but were resistant to change in work routine (not how we've always done it) and slow to adopt new technology. The younger generations embrace change, have a higher tolerance for risk taking, make decisions based on their near-instant access to information, and have absolutely no fear of technology because they've grown up with it. Rather than work the routine, they seek to improve it. Make it faster, less wasteful, more efficient.

Scott Herod, senior program manager at North American Stamping Group (NASG), believes that relationship building between the generations in the workplace breeds confidence and trust, leading to transparency and openness, and helps overcome the natural road-blocks to collaboration.

"When a 55- to 60-year-old worker is paired with a 20-year-old new hire," he says, "it's more natural for the younger worker to feel that they're learning from the older one. The inverse seems unnatural, but that's what needs to happen."

This is where leadership needs to push commonality and under-standing by showing the benefits of both experience and innovation. In manufacturing, the older generations are already predisposed to continuous improvement in operations. They're familiar with the concepts of lean manufacturing and having a Kaizen culture around specific processes to improve throughput and reduce waste. What they're uncomfortable with is the speed of technological change in the plant and what that means for them jobwise. Each generation compares itself to the one coming behind them. Do they have the same work ethic, ingenuity, industriousness, dedication, etc.? They look for differences rather than common ground.

Shane Vanderkerkhoff, Director of Client Development & Solutions at RJG, pairs young and old workers to encourage understanding and overcome the reluctance of older workers to step outside their comfort zones.

"We start by challenging convention," Shane tells us. "Why did we do it this way? When was the last time we analyzed this process? Do we have better tools today? We encourage them to question each other about reasons why, explain and discuss the benefits of the new technology or process, and then reiterate the end goal to reach a consensus. We have them find out what they agree on and ask them to start there."

As a leader, Shane is looking for reactions to and understanding of needs, wants, and strengths.

"In intentional cross-functional team-building, the team must have an appreciation for individual strengths and how to best apply them to bridge perceived gap and for the good of the company," he says. "Two-way communication is key."

Jenn Barlund at Falcon Plastics believes that creating a culture to drive change is at the heart of it. She should know. She's young, in a leadership role, and has encountered a fair amount of skepticism regarding her ability to lead a manufacturing workforce. She has had to influence and engage people who are much older and more experienced. She acknowledges the differences between generations but stresses that understanding how people are wired and how they think are crucial.

"You have to be able to have conversations that revolve around shared values, current realities, common ground, and the information that is in front of all of you," she says. "Just because we've always done it this way doesn't mean it can't be done better or more efficiently. We look at processes for optimization and at new tools and new technology that can contribute to efficiency and demonstrative success. We do this with consistency, clarity, and communication of expectations. It's not easy, but it moves the needle."

Glenn Whitecotton, Value Stream Leader at Hansen Plastics, also acknowledges the differences between legacy and next-generation leaders. He describes younger leaders as being more vocal, excited to share ideas, but willing to move on if they feel they're not being heard. By contrast, legacy leaders typically came from a background of simply doing what they were told so they're slow to adopt and adverse to change. He believes that the key to persuading older employees to step outside their comfort zones to embrace new technology is to show them the benefits.

"I often find that if someone is convinced something will make their job 'easier' they'll work to ensure its success," Glenn says. "I try to gain their input, include them from the outset, ensure that their voices are heard, and challenge their conventional thinking in a respectful manner."

Grant Johns, President of PolySource, feels that preserving tribal knowledge is paramount.

"Legacy or tribal knowledge is leaving faster than we can transfer that knowledge to the next generation, and we're not bringing enough new talent into the industry to replace who's retiring. We're losing a lot of great people. So, we need to be intentional with time overlap. We need the generations in our workforce to work together so they can learn, ask questions and benefit from a kind of fluid knowledge transfer."

There is resistance, however, and a basic difference in work ethics. The younger generations value personal time more than the older generations. They work to live rather than live to work. They trust technology because they grew up with it. The older generations feel forced to adapt to it and distrusts it because it's so unfamiliar. Grant also feels that making sure that empathy, trust, and understanding play a large role in bringing the generations together for the good of the company.

Grant offers an example of how he dealt with a co-worker resistant to change:

"A colleague of mine is a technical leader in R&D whose work resembles science projects more often than not. R&D is rarely an efficient process and he'll put together complex formulations with lots of moving parts and constituents. When we deployed a new order management system at PolySource, he was reluctant to adopt it and instead, created his own orders outside both the system and the Order Management team. When we asked him to work with the team and the system, he simply didn't want to and had a list of reasons why mostly concerning accuracy and complexity. Our problem was that his behavior was creating problems on the back end. So, we came up with a suggestion that he work with the most detail-oriented member of the OM team, a young woman who happened to be phenomenal at catching discrepancies. I told him 'I need you to trust me and take this risk because I believe it will pull a lot of time out of your process.' He tried it, reluctantly, saying 'You're determined to make me change.' I told him that I was just trying to talk him into making the journey. In the end he came around, adopted the new system and said, 'Don't ever make me go back.'"

"It's important that all of our people understand the 'why' in all our activities," he says. "Why are we asking you to do this, make this change, employ this new technology. It requires clarification of our common goals and builds trust throughout the organization. Leaders have to be able to trust their teams, but they also need to understand why people feel the way they do, why they may be resisting change, or why they're eager to embrace it."

All of these leaders have different perspectives to engage the experience of the previous generations to the efficiency of the new generations. There is no single answer to this challenge, but the acknowledgement that it is critical to learn from the legacy employees and combine that with the excitement and efficiency of the new employees to allow companies to move to the next level of flexibility.

SUCCESSION AND SUSTAINABILITY: TWO SIDES OF THE SAME COIN

Of all the small- to medium-sized North American manufacturing companies we have personally studied, a significant percentage of them are either owned or led by an aging population, and approximately half of them do not have succession plans. This means that they don't have a younger family member who is qualified to step into executive management or wants to, or more likely they haven't yet identified a suitable candidate from inside or outside their organization. Many of these older leaders have always expected the next generation or their grandchildren to come up through the business to continue the family legacy. Unfortunately, many of these next-generation family members simply aren't interested or aren't at all sure that's what they want to do with their lives.

For legacy leadership, succession planning is a delicate balance. They have to be willing to let go, to loosen their grip on the reins. They need to admit to themselves and the company that things will change. Operations may be different from how they'd run them. Succession planning is, after all, the evolution of the business. It's also a fact that not all of an owning family's members are cut out for leadership roles. An astute legacy leader with the welfare of the company at heart should be willing to look outside the family for leadership talent or next-generation development, when necessary, but also consider company insiders for future leadership roles. Those candidates who are actively demonstrating those traits that play to the company's strengths are the ones who can shepherd the business to the next level. They understand the long-range challenges of the industry. They care for the company and are transparent and non-judgmental. They question, consider, and accept with humility. They collaborate in creative thinking and problem-solving, and they have the innate ability to influence without exerting authority. They know and appreciate the difference between strategy and tactics and have the ability to both view the big picture and break it down into workable components.

For young incoming leaders it's imperative they realize that they don't know everything. They need to actively seek to understand the things they are not familiar with in the company's operations and be aware of the team's apprehensions regarding management change by asking sensitive questions. They must leave any sense of entitlement, arrogance, and pushiness (I'm the boss) at the door and assume a leadership role with humility. Remember that leaders are the stewards of the company's culture and values. They must recognize what they're good at and leverage it. It's not necessary to try to be everything their predecessor was. Instead they need to gather good, capable people who don't necessarily share the same opinions but are catalysts for respectful critical thinking. Even leaders need mentors, coaches, advocates, and advisors. One individual can't possibly have all the answers.

Many owners are simply not sure that they're ready, much less willing, to turn over the keys to the kingdom, particularly to an outsider or to the youth. Readiness is a two-way street. Incumbent leadership has to be ready to let go and incoming leadership has to be up to the challenge at every turn. They should expect to stumble yet not fear failure. Missteps are an important part of the learning process and how the young leader reacts, remedies, and moves on will be watched closely by everyone. No leader, experienced or newly promoted, has all the answers. We all learn from our failures!

Succession planning, however, shouldn't be restricted to the corner office. It's a much larger issue and involves the entire company from the loading dock and the shop floor up to executive management. It's not just about who's going to be the next president or general manager. It's also about who's going to head up maintenance, engineering, HR, who's supervising on the shop floor and who's sweeping it. It's about planning ahead, securing the future for the company, and then sustaining it. The recent disruption of the Covid pandemic drove this point home for many manufacturers who had to quickly decide what to do when entire departments or strata of senior management were sent home or worse, contracted the virus. Remember the lottery winners we mentioned in a previous chapter? An entire engineering department played and won the lottery and then quit as instant millionaires are likely to do. That's an extreme example but the fact remains that there were no plans in place to handle the sudden absence of an entire discipline.

WHO'S NEXT IN LINE?

If current leadership is not already grooming a key leader or family member to step into the top spot, how is the next leaders identified? Look first in the company's own backyard. A good succession plan is a combination of internal and external talent. Chances are good that the expertise and drive needed in new leadership is right under your nose. Home-grown talent knows your manufacturing sector, your processes, your customers, and competitors. They are diamonds in the rough, already culturally aligned and invested in the success of the company. Look for team members who are not intimidated by challenges or being challenged, who personify and live/work your company's values, and who have demonstrated critical thinking at both strategic and tactical levels.

How NOT to Succession Plan

A hypothetical example: Let's say that a manufacturer acquires a smaller company and, as part of the assimilation/consolidation process, evaluates every plant manager at the acquiree. After review, managers who didn't have higher level degree, but rose through the ranks are purged. The result is disastrous for the manufacturer because by ejecting managers based on a blanket educational standard rather than hard-won experience and proven performance criteria, the acquiring company has effectively thrown out all that tribal knowledge and the ability to run well. All organizations have those individuals that just know how to get things done!

Succession planning is also tightly tied to the issues of generational knowledge transfer that we discussed earlier in this chapter. If one or several individuals are the keepers of a company's tribal knowledge—the "Yodas" who remember customer histories, why processes were engineered the way they were, and the critical details of specific challenges met and overcome through the years—and there is little or no effort to capture and preserve this knowledge, then much of the historical critical thinking that contributed to the company's success is lost. When the lessons of the past are left in the past, success is that much harder to sustain going forward. Remember what history might have taught us.

Of course, succession planning and sustainability come with their own set of potential barriers that relate to both the culture of the company and the generational issues we've already touched upon. How do you get past or get through to employees who guard their roles and responsibilities and refuse to share their knowledge and expertise for fear of becoming expendable? To counteract this tendency to protect and hoard their knowledge and capability, work to move them up the accountability ladder and help them understand that they are standing in their own way—a barrier to their own success. But first you have to convince them that their behavior affects the whole company and encourage them to talk about what makes them uncomfortable to share what they've learned.

This is an opportunity for leadership to make sure *everyone* realizes that they are part of a unified organization, a team that works for the good of the whole, not just the individual. As a leader, you have to be willing to force the dissemination of captive knowledge and expertise. The most effective leaders are not afraid of making the tough calls, the hard decisions, because they are consistent in their behavior, consistent, transparent and foster the free-flow of ideas, concerns, and discussion at all organizational levels.

The phrase "the whole is greater than the sum of its parts" comes to mind because it's true. No single employee is solely responsible for the success of an organization. We all are. Succession and sustainability come from culture, understanding, and the timely development of primary and backup plans covering known contingencies as well as unexpected events and conditions. And when carefully designed and implemented make our companies stronger and more resilient—ready to face whatever the future brings.

Consistency and Firmness Builds Trust

If your people don't see you consistently and firmly handling issues in a timely manner, you run the risk of losing your team's trust and respect. Situations or conditions that detract from sustained, healthy operations such as difficult employees, negative influence and/or challenging business scenarios are "energy vampires." At Falcon Plastics, Jenn Barlund handles what she calls "less than optimum" situations by actively de-escalating the mood and lowering the temperature of discourse. She works to take personal emotion out of it. "Drama doesn't solve problems," she says. "It tends to make things worse." Remember, your people are watching you. Consistency and transparency—not fairness, which is in the eye of the beholder—are foundational when it comes to building sustainable success.

Many leadership books have been written to help leaders take their businesses to the next level. We are not sure that we have anything else new or revolutionary. The key to leadership is humility, vulnerability, and consistency. The best leaders have acknowledged that it takes a village and they don't have to be the smartest. One of our newest customers in 2022 told us to be honest with him and tell him if he is not the guy because he will step out of the way and allow someone else to take the company to the next level. The truth is, he is the guy, humble, willing to be taught and coached and is already moving the company to greatness.

HAVE YOU CONSIDERED?

1. Is your company ready for new leadership?

2. Do you acknowledge the efforts and achievements of your workforce?

3. Are your workers comfortable asking you for feedback?

4. How would you characterize the leadership style in your company?

5. Are your managers authoritarians or influencers?

6. Are you organized and prepared to react to unforeseen events in your market?

7. Can you describe your company's culture and values? Are they clearly communicated?

8. Are you comfortable admitting that you don't know the answer to a question from a peer or member of management?

9. What do you consider to be the most important leadership traits?

10. Have you ever analyzed your own work habits or considered what accidental messages you may be sending with your workplace behaviors?

11. How do you handle team members who think they know it all?

12. When implementing initiatives do you establish ground rules and clearly communicate expectations?

13. Are your people reluctant to raise issues or voice their concerns?

14. When a worker brings a problem to you, do you empower or enable them?

15. Where do you think you are on the accountability ladder? Where is your team?

16. How do you react when something goes wrong?

17. Do you have a knowledge transfer plan?

18. Do you encourage collaboration between generations in your workplace?

19. What are you doing to capture legacy expertise (know-how)?

20. How do you convince a worker or colleague to step outside their comfort zone?

21. Do you know who in your organization is ready to lead?

22. Have you identified the up-and-comers in your organization? What are you doing to encourage them?

23. Are you willing to make the hard calls and tough decisions, or would you rather push the responsibility onto someone else?

24. As a family member to the founder are you humble and willing to hear feedback regardless of your role in the company and/ or family?

THE HUB OF
THE WHEEL

Throughout our discussion of the existential need for transformation and generational improvement in North American manufacturing, we've touched on the five areas of focus that impact the success of most manufacturing enterprises—operations, labor, commercial, vision and strategy, and technology. At the center is Leadership.

Leadership determines direction and promotes and enables action. It motivates. It guides. It sets the standards, the mood, the culture, the expectations, and the objectives. Without it, sustained success is simply not possible. And it's the strength of the core or the hub of this wheel of opportunity that holds it all together. People want to follow good leaders!

In chapter 7, we explained why it's so important that the industry evolve leadership to keep pace with competitive, economic, and technological change and what that will likely entail. We talked about the pressing need to capture the decades of practical expertise and manufacturing know-how that's flowing out the door with retirements. We discussed the importance of combining this tribal knowledge with the enthusiasm, new technologies, and practices of the incoming generations of digital natives.

The next-generation leaders can and will take North American manufacturing to the next level of global success. They will have both the inclination and obligation to challenge the norms, hold their leadership peers and their teams accountable, anticipate and adopt the next best thing. We'll see positive conflict, uncomfortable at times, but necessary to move forward not in incremental steps but by leaps and bounds, navigating step changes, shifting paradigms, and busting conventional thinking.

In the early pages of this book, we described our industry as at a tipping point. If we don't encourage the next generation to enter the manufacturing sector by making it a better, more attractive place to work with a sustainable future, then the industry as we know it will dwindle away. The people we need to influence, aside from next gen workers, are the parents and school counselors and demonstrate to them that manufacturing is not a dinosaur industry, but instead offers worthy, technologically focused and enhanced careers. And then we need to convince the retiring generation that leaving a legacy is germane to the future of the industry.

North American manufacturing is a very vulnerable space right now. If we do not bring in the next generation of leaders and pass down that legacy, the industry will disappear. And if we don't encourage those new leaders into manufacturing by making it a great place to work where they can make an impact, then sooner or later the industry will still disappear. It has to start with schools, counselors, parents—we have to show them that manufacturing, especially in the rise of technology, is a worthy and successful career option. And this shift has to be led by the retiring generation, leaving their legacy for the future of manufacturing in this country.

With the shift change in leadership and the proactive initiatives and practices outlined here, we're confident that it can be tipped in our favor, and that North American manufacturing can and will improve, recover, and regain the forward-thinking innovation and can-do attitude that once defined this proud industry.

The time is now. If we don't shift our manufacturing mindset and embrace the potential of the next generation of innovators and leaders, the global economy and ever-changing world will push the survival of North American manufacturing out of reach.

LAURIE HARBOUR

As president and CEO of Harbour Results, Inc. (HRI), Laurie Harbour leads a team of analysts and manufacturing consultants to help small- to medium-sized manufacturers develop short- and long-term strategies, improve their operations, reduce risks, and optimize business.

Driven by her passion for manufacturing which developed at a young age when her father took her weekly to the Chrysler plant in Hamtramck, MI, she has pushed for more woman and diversity in manufacturing. Laurie told her father at twelve that she wanted to be in manufacturing, and he told her no, that wasn't possible—it was too hard and she was a girl. That fueled a fire inside of her to make an impact in manufacturing for the rest of her career. Spending most of her life around the manufacturing industry and later being mentored by her father Jim Harbour, she utilized her experience and knowledge to found HRI in 2005. Since that time, she partnered with Scott Walton in 2007 and further grew in her manufacturing capability. Laurie, along with Scott, has been responsible for developing both the consulting services and targeted tools to profitably grow the organization into a leading manufacturing consulting firm.

As a trusted advisor to the North American manufacturing industry, HRI monitors, researches, and analyzes the manufacturing value stream to identify strengths and weaknesses, gaps and risks, and business and operational opportunities in an effort to help the industry transform to be more successful in the global marketplace. Additionally, the company is the leading forecaster for the automotive tool and die industry, collecting and analyzing data on a regular basis through Harbour IQ, a proprietary market intelligence tool that collects thousands of data points from hundreds of manufacturing companies annually. Laurie and her team utilize this data along with their expertise to help companies solve problems, improve business, and advocate on behalf of the manufacturing industry. As an industry-thought leader, she is regularly quoted in both business and trade magazines. Additionally, she speaks to companies and at forums on topics such leadership, cultural transformation, supplier collaboration, manufacturing, overall business operations improvement, and manufacturing forecasting. In 2018, she was named *Crain's* "Notable Women in Manufacturing" and *Plastics News'* "Women Breaking the Mold." Additionally, she was named the *Plastics News* "Automotive Newsmaker of the Year" for 2018.

SCOTT WALTON

Scott Walton is chief operating officer and chief financial officer at Harbour Results, Inc. With more than thirty years of experience in strategic planning, operations management, lean manufacturing, and supply chain management, he has assisted multi-national companies, government entities, and business professionals worldwide. Walton's expertise lies in automotive, plastics, defense, and distribution. He works closely with automotive, manufacturing, and distribution companies to assess their business needs on a strategic, financial, and operational level. Walton is also an expert at restructuring manufacturing companies with a focus on quickly diagnosing key issues and developing plans to restructure finance, sales, and particularly the operations of the company. Scott oversees the Harbour team of senior consultants that assists clients in implementing operational and strategic changes, driving throughput, and improving profitability.

Prior to joining HRI, Walton was president of the automotive market segment for Nypro Inc., a global leader in custom plastic injection molding where he was responsible for the development and deployment of auto strategy on a worldwide scale. In addition, he held operational responsibility within four automotive-focus factories (two in the United States and two in Mexico) and a sales/engineering office in Detroit. In addition to his work at Nypro, Walton has held various leadership and operational roles both domestically and internationally. He also is a skilled communicator and frequently presents at manufacturing industry forums and events as well as directly to manufacturing companies.

Together Laurie and Scott have built a niche consulting firm based on the use of data and market intelligence to help companies understand where they stack up and how to drive improvement quickly. Their passion for manufacturing is evident in all that they do and their commitment to their clients is second to none. They get out of bed every day to make an impact in the lives of all they encounter daily and to make a difference in North American manufacturing.

GET IN TOUCH

Harbour Results, Inc. (HRI) is a trusted advisor to the manufacturing industry. Established to help small- to medium-sized businesses improve, we provide operations improvement, strategic development, custom analysis, and market studies, benchmarking, and business assessments. HRI combines data with knowledge and expertise to deliver customized strategies, implement operational improvement, and drive sustained business success.

As leaders of the HRI, Laurie Harbour and Scott Walton are experienced manufacturing leadership speakers and coaches. Each with more than thirty years of manufacturing experience, they have participated as keynote speakers, hosted workshops, and conducted small-group training sessions. As authorities in the manufacturing industry, they are driven by a passion to revitalize the North American manufacturing industry.

Harbourresults.com

For more information on benchmarking or consulting, please email *hriwebservices@harbourresults.com*.

For more information on keynote speaking and coaching, visit leadersinmfg.com or reach out to *laurie.harbour@ harbourresults.com* or *swalton@harbourresults.com*.